Honey

Honey

Recipes from a Beekeeper's Kitchen

AMY NEWSOME

Photography by Kim Lightbody

Hardie Grant

QUADRILLE

006	Introduction
010	A Beekeeper's Year
024	Cooking with Honey
034	Honey Terroir
044	Restocking Nature's Larder
060	Chapter 1 – Jars
076	Chapter 2 – Small Plates
104	Chapter 3 – Large Plates
138	Chapter 4 – Bakes
212	Chapter 5 – Ices
226	Chapter 6 – Drinks
244	Index
252	Thank You
253	Author Biography
254	Cook's Notes

Introduction

Honey has been a fundamental part of human existence for over 10,000 years. From Stone Age wild honey hunts to ancient Egyptian hieroglyphs depicting beekeepers using clay pipe hives, the magic of honey flows through the past and present of a myriad of cultures the world over.

We spoon it into tea, drizzle it over yogurt and spread it on toast, yet if we look beyond the bland squeezy bottles on our supermarket shelves to raw, unprocessed honey, an untapped world of flavour presents itself.

Honeybees forage nectar and pollen from flowering plants within a mile or two of their hive. Each plant species in nature's larder gifts honey with a specific taste, colour, aroma and viscosity. The beekeeper works in perfect rhythm with the season's nectar flow, knowing when each blossom bursts and fades; their relationship to the local landscape rivals any French vigneron. A spring English honey might be pale gold and creamy, thanks to bramble and hawthorn blossom, while in autumn it may be dark, runny and rich, reflecting plants like ivy and rosebay willowherb. An Australian winter honey might be verdant and aromatic from Eucalyptus blossom. Delightfully, no two honeys are the same.

I've written this book to lift the veil a little on the mercurial craft of beekeeping and cooking with honey. We'll follow the bee's wing through the seasons as different flower species bloom, offering a fresh insight into how the best honey is made, and why single-origin honey from small-scale beekeepers is the ultimate flavour expression of terroir; just like the best wine, extra-virgin olive oil, coffee and chocolate. There's ample advice for bee-friendly gardening, learned in my work as a Kew Gardens-trained horticulturist and garden designer, and an in-depth look at the marvel of pollination, to discover just how flowers and their insect pollinators work together to produce the pot of gold in your kitchen. Honey has inspired generations of cooks to dream up delicious dishes and drinks both savoury and sweet; here curious cooks will find new takes on familiar favourites, much-loved historic traditions and modern plates.

Quite often I get asked, 'but which are you really: a beekeeper, a gardener, or a cook?', anxiously looking to put my life into the 'correct' genre and find out which one is 'just' a hobby. I'm none without the others, so I'm afraid it's all three. We are never wholly one thing, I think.

Each practice inspires the other, and doing one restores you before returning to the other. You might be a parent, but also a software engineer and a keen cyclist. You might be a part-time baker, and a part-time writer, but fully devoted to both.

This book is a love letter from a beekeeper, gardener and home cook to good food, good honey, and the bees that make it. I've been beekeeping on a small scale for six years, which is not long at all for one of the world's oldest professions. So this is not the definitive, encyclopaedic book on honey or beekeeping, far from it, but a way in to excite curious minds, sharing what I do know, and giving you a glimpse into the magical world of the bees. Beekeeping in the UK is often associated with the older generations, and fresh local honey with traditional farmhouse recipes. Not so here.

I live in London, and the amazing creations of this city's talented chefs, bakers, bartenders and restaurateurs make my heart sing, so naturally you'll find their influences scattered through these pages, as are some more lively ingredients than you might expect: black lime, guajillo chilli, and plenty of home smoking, as I like to cook over fire. I am a home cook, not a professionally trained chef or food anthropologist, and I am all too aware of how much I don't yet know, to do justice to the full diversity of the amazing honeys, dishes, and related cultural practices around the world. But this is the book that I can write at this moment in time. What matters is being curious to learn more. I hope these words will make you curious about honey, bees, gardening, and, above all, itching to get started in the kitchen.

A Beekeeper's Year

Spring

March

The ghostly stillness of the winter apiary melts away in March, as crocus, winter honeysuckle and wood anemone come into flower. Temperatures begin to rise (and invariably falter, frosting our magnolias) and longer sunlight hours wrest the hives out of hibernation. Impatient, I loiter, casting an eye out for the first whisperings of movement at the hive entrances, signalling successful overwintering. Inside, the bees are unfurling from their winter cluster, and begin to vibrate their wings more intensely, raising the hive temperature to a nurturing 35°C (95°F), ready for the queen to resume laying eggs resembling 2mm-long rice grains, at a rate of up to 2,000 eggs a day, to swell colony numbers ready for spring foraging.

On days when the air temperature reaches around 14°C (57°F), worker bees venture out to forage on the earliest spring blooms close to the hive; here in the south of England it is blackthorn, crab apple, willow and wild cherry blossom. It is usually still too cold to open the hive up to the elements and take a proper look; however, the beekeeper can get away with a quick 'heft'; lifting the bottom of the brood box slightly to judge the weight. If it feels light, the bees may be low on honey and need feeding, as there may not yet be enough forage flowering, or high enough temperatures to fly, or enough foraging workers yet inside the hive to bring it home. As through winter, when this is the case, we feed them either with their own honey if we have it stored, or with sugar fondant if they didn't make enough honey the previous season.

April – a first glimpse

The first spring opening of the hive is always a heart-in-the-mouth moment. Smaller, weaker colonies may not have made it through the winter months, and the sight of a perished cluster of motionless bees is a profound lump in your throat never forgotten. With temperatures still a little bracing, this first hive check is focused and brief to conserve the bees' warmth. I am impatient to see how they're doing.

As you would at home when it's chilly, the bees seal every draught and gap of the hive's woodwork with propolis, a mixture of foraged tree resin (usually poplar) and beeswax, which glues everything solid; the name propolis comes from the Greek *pro* 'at the entrance of' and *polis*, for 'community' or 'city'. The beekeeper's hive tool is the key to the city; a deft lever, wedging open each layer of the hive with a satisfyingly crisp 'crack'.

With my mind on the clock to avoid chilling the bees, I remove the roof and work down towards the brood chamber, prising a frame up from its sticky propolis cradle for a quick look. The bees are slow; I spot the queen meandering through the crowds, looking for the next hexagonal cell to lay an egg in. Cautiously noting that all seems well for now and mindful of the cold, the temptation to have a more thorough look through is resisted. I quickly put the hive back together again to keep them warm.

True overwintering success comes in milestones for the surviving hives. The queen must begin to lay eggs again and rapidly, ensuring there are enough new worker bees to take over from the spent winter population, while old and new workers must bring in sufficient forage to support such a baby boom. It's a delicate balance. The inevitable April showers make things even more precarious; bees can't fly in the rain, so they must sit it out, prevented from bringing in more forage, while using up the last of their winter stores. It's a nail-biting time for a beekeeper, and I'm often found worrying around the apiary or mixing up sugar syrup in the kitchen as a last resort bee feed if stores are low or the weather forecast is poor.

Much like the bees in their hives, beginner beekeepers have been cooped up in the classroom over winter, studying the theory of the craft and itching to open a hive. As April warms, it is finally time. Here in the UK, once the flowering currant (*Ribes sanguineum*) blooms, which is usually around 17°C (62°F), it's time to begin full hive inspections every 7–10 days. The clocks have 'sprung' forward, the evenings are lighter, and it's a joy to spend an early evening getting stuck in at the apiary.

Under the watchful eye of an experienced mentor, very slowly you remove the hive roof and the crownboard for the first time. No amount of theory training can prepare you for the feeling of finally seeing inside a live hive and marvelling at the rumbling mass of busy bodies below. For some people this is a lively stick-or-split reckoning. Despite best intentions, the buzz of a trail of bees popping up close to say hello (or more accurately, to ask WHO are YOU?) can inspire a maniacal run to the garden gate. Happily for many others, the thrill of spotting the queen, watching the worker bees go about their business and 'getting your eye in' to make out the blindingly tiny eggs is the start of a glorious apiological adventure.

Hive 4: Thursday 14 April, 4pm

17°C, sunny, slight breeze

Notes:

First full inspection after winter.

Busy activity at hive entrance, bringing in plenty of pollen.

Overwintered on double brood, no queen excluder, one super.

Relaxed temperament. Super empty.

Top brood box: Queen, blue dot. 8 frames of brood in all stages, plenty of pollen in various shades. No visible varroa or wax moth. Removed wild comb from end of brood box. Old nuc frames already moved to outer edges of cluster.

Lower brood box: Mostly old frames, very dark comb. Empty, no eggs, brood or food. Outer comb furthest from colony moulded.

Work done: Removed lower brood box and combs disposed. Added queen excluder, and pack of fondant on top, inside super with central frames removed. Swapped out stand and lifts for freshly painted set. Strimmed grass around and under hive. Removed entrance reducers.

For next time: Keep an eye on food levels – not bringing in any nectar yet and super was empty. Mix up spring syrup for feeding if necessary. Make up super and second brood box with frames, ready for honey flower and returning to double brood. Add varroa board.

In flower: Daffodil, weeping willow, tulips, pear blossom, hawthorn, medlar, quince.

May

There are three types of honeybee inside each hive: the queen bee, a seasonal sprinkling of male drone bees, and lots of female worker bees, which make up around 90% of the hive population. As the name suggests, they work hard; once a worker bee is old enough to forage, she leaves the hive in search of nectar or pollen to feed the hive, passing through gardens and hedgerows. As a bee passes from flower to flower pollinating our plants, its 'fur' becomes dusted with pollen, which the bee combs off into pollen baskets on each hind leg to take home. Different plants produce pollen with different colours, drawing a tapestried map of the bees' foraging journey. The deep orange of dandelion jostles cheek by jowl with the pale gold of rosebay willowherb or the pure black of oriental poppy pollen.

The queen bee is much larger than either the worker bees or the male drones, and you can identify her from her much longer, slim and smooth abdomen, which extends past the end of her closed wings. In a managed beehive, she's also easy to spot because of a dot of bright paint on her thorax. This is put there by the beekeeper or bee farmer to allow easier queen spotting and show the age of the queen, with one of five colours corresponding to the last number of a year; a queen bee born in 2023 would have a red dot on her back.

YEAR ENDING IN	COLOUR	MNEMONIC
1 or 6	white	will
2 or 7	yellow	you
3 or 8	red	rear
4 or 9	green	good
5 or 0	blue	bees

We look for the queen during hive inspections to make sure the colony is 'queenright', because if the queen is present and laying eggs as expected, the colony is usually doing fairly well. During swarm season from May until September, the hive could be looking to swarm, which starts by them raising a new second queen. As the queen is very large, she requires a bespoke cell, built as an extension from a regular hexagonal cell, which when finished looks knobbly and strange, like the shell of a peanut. Perfection takes practice for these highly specialised cells, so worker bees make 'play cups', beeswax acorn cups, as they learn and begin to think about swarming. Each inspection involves looking for these cups, seeing if any are being extended and 'charged' – filled with an egg and some highly nutritious royal jelly – then a decision must be made. At this point the beekeeper can make an entirely new colony 'for free' by artificially swarming the bees. The frame with a charged queen cell is placed into another beehive, along with a couple of frames of bees, brood and food. With luck, she will hatch and the colony will expand out to fill the new brood box. As a new queen, she will need marking with her own spot of paint, which is a very careful process involving the cutest of unique tools fit for Thumbelina – a tiny queen cage, a tiny pot of paint, the end of a matchstick, and a very gentle touch.

Hive 3: Thursday 12 May, 4pm

19°C (66°F), cloudy, still

<u>Notes</u>: Massive! Bursting at the seams with bees. Good temperament.

A few Q cups, 1 charged. Removed.

Saw the queen, and BIAS (Brood in All Stages)

Wild brace comb full of honey in super, where they'd raced through a small pack of fondant and filled the space. Whoops. Left this for the bees as nectar flow has been limited.

Spotted a couple of varroa mites.

<u>Work done</u>: Added extra brood box, super and two lifts. Put in inspection floors to count mite drop over the week.

<u>For next time</u>: Check varroa mite drop count. Replace brace comb with super frames.

<u>In flower</u>: Dandelion, iris, wisteria, red campion, weeping willow, tulips, apple blossom.

Summer

June

'A swarm in May is worth a load of hay, a swarm in June is worth a silver spoon, but a swarm in July isn't worth a fly.'

This 17th century saying still rings true today. In June, beekeepers can often be spotted rootling in a hedgerow or chimney, dashing around with a bedsheet and a cardboard box, catching their own swarms or helping to take care of others'. A late spring/early summer swarm has time to settle and grow into a good-sized colony before the following winter, whereas a late summer swarm may not make it through on its own.

Swarming is an incredible natural phenomenon to be lucky enough to witness. It is the reproduction of a superorganism, as the colony splits into two, raising a new queen to stay in the original hive with the younger worker generations, while the old queen departs with the mature workers for pastures new. A clump of worker bees will settle around the queen on a branch, table leg, chimney stack or garden chair, as scout bees zoom off in search of a more suitable permanent home. It's a striking sight but nothing to fear, as swarms are surprisingly calm. A clustered swarm is scoopable, like a thick mousse, and can be dolloped by hand into a box, ready to be 'hived'.

Wednesday 15 June, 3pm

Swarm!

Notes: Heard a mild roaring to my right as I walked down the garden path this afternoon; a large cloud of bees were assembling above the apiary. Hive 1 is swarming. I kept my distance and got on with gardening while they decided where to go.

Half an hour later, they'd settled in a clump on a tree branch nearby, luckily at head height. I prepared a new hive with frames to take them, and gathered an old poly nuc box, secateurs, and a bee brush ready.

I put my suit on and lit my smoker just in case. Holding the nuc box closely under the cluster of bees, I cut the branch, so that it gently dropped into the box. I carefully pulled the branch out, lightly brushing off any hitch-hikers.

Hived swarm and added some honey from their old hive. Fingers crossed they take to their new home.

Work to be done: Check on them in a week, and check the old hive for a new queen.

In flower: Blackberry, clover, sweet chestnut, buddleia.

For honey-minded beekeepers, swarms are tricky, because this regeneration means losing all your nectar-foraging bees and your reliable queen, then waiting for the younger ones to grow up and start foraging again, while the new queen must mate before getting up to speed on egg laying. All this palaver halts the honey flow for several weeks, and there are often more swarms to follow, so many beekeepers try to avoid this, from pre-emptively splitting a colony in two, to making the bees think they've swarmed when they haven't, to adding another brood box in the hope that a generous house extension will be enough of a temptation to stay put. We are still learning what makes bees swarm, but it is a natural impulse, so to control it always feels discordant. While the beekeeping practices described in this book are part of the 'traditional' approach here in the UK, there are many different ways to look after honeybees, more or less hands-on, some swarm preventative or not.

July

The relationship between a beekeeper and their bees ebbs and flows throughout the seasons. In summer, we surge together. Sun-warmed wood and heavy frames of honeycomb are lumbered to and fro, brushing through long grass and ox-eye daisies. Sticky with sweat and honey, the air thick with buzzing wings and laboured breath. Clouds of enlivened worker bees surge up from the first opening crack of a hive inspection to my veil to say hello, and half-heartedly chase me back to the shed afterwards.

If you are lucky, you'll barely be able to keep up with the honey flow when the bees are in full summer swing; the hive will grow taller and taller with each new super (honey box) you add, making sure there are enough shelves in the hive pantry for the relentless incoming of golden nectar. It is quite miraculous to watch each box of empty frames fill out and up.

Worker bees produce beeswax, which is secreted through glands in their abdomen as large wax scales – exactly like, umm, dandruff – to be chewed and moulded into the perfect hexagonal cells, quickly building out the comb on both sides of the sheet. This process can take only a few days in peak summer; after adding a new super of empty frames on a Monday it could be fully 'drawn out' (hexagonal cells built) and filled up with honey by the following Monday.

August

For migratory heather beekeeping, an ancient practice, late summer is time for a hasty hoisting of anchors. Hives are ratcheted together, piled on the back of a truck at dawn and transported right into the heart of the purple heath to catch the bell heather nectar before the blooms fade as quickly as they blossomed, sometimes only for six days. Heather honey is unique for being 'thixotropic', which in itself is a uniquely wonderful word. This means that it becomes runny with movement, like the stir of a spoon, but once still again, it returns to a set jelly-like consistency. Its mercurial nature also extends to its microbial activity; heather honey is viewed as a match for manuka for its health properties. For me it's worthy of the hype for its flavour, which is richly aromatic, deep yet tart and resinous. I love it in cream-based puddings, like panna cotta or a golden custard tart.

Wednesday 9–10 August:

Helping Simon move his bees to the heather.

Spent a couple of days in the New Forest with beekeeper Simon Noble and his apprentice Chloe, learning a little about how they do things, and helping them move their bees up to the heather. Simon has 70 hives, and a special permit from the New Forest National Park to move a select few onto the heather when it blooms for around six weeks in late July/August.

We meet in his forest apiary at 5.30am, closing up the hive entrances with pieces of foam to keep the bees inside, and use wheelbarrows to gently load them into a van.

We bundle in together for the sunrise drive through ancient woodland and out onto the open heathland, with flasks of tea between our legs. A special key from the park authority is used to access a more remote area away from the main roads, to find the perfect quiet spot for the bees.

We unload the hives in the dewy bracken, before the sun comes over the tree line. Trying not to trip over fallen trees or into cattle dung, we line them up on the edge of the forest, looking out over the heather. Time for a cup of tea, soaking up the view as the bees settle into their new surroundings. With the intense heatwaves and drought, Simon is sceptical that the heather will provide much nectar flow this year. Because of heather honey's jelly-like properties, it is difficult to extract – Simon uses an apple press. It costs more for the extra time, effort and skill it requires to make, so it's an important honey for Simon's business, as well as an ancient tradition to be preserved and kept alive. Even if there isn't much to sell this year, Simon takes comfort knowing that what little there is to be had will be enjoyed by the bees first, as he only takes what isn't needed.

Simon calls us to follow as he darts off into the trees nearby, checking on a wild colony he saw last year who have taken up home in the trunk of an ancient yew tree. We peer through the branches, mugs still in hand, tea slopped over our beekeeping suits, watching the bees come and go from a hollow in the trunk.

We pack up the van and return to Simon's honey HQ, where Chloe and I spend a couple of hours uncapping and spinning honey frames, and jarring up fresh batches from an apiary in a local meadow. We head to the coast for a lunchtime dip, and I make my way back to London in the afternoon haze.

Autumn

September, October – seasons of honey

Beekeepers harvest honey at different times for different reasons. If you have a keen eye for the local bee forage and watch carefully for when different plants come into bloom, you can put on an empty super just as blackberry comes into flower, and remove it once the blooms start to fade, in order to harvest blackberry blossom honey. In practice, unless your apiary is placed within a large monoculture like a many-acre orchard or fields of oilseed rape or lavender, it is hard to be sure of where your bees have been. The honey crop is harvested through summer into September, leaving enough behind for the bees to feed on over winter, and enough time through September and October to bring in the very last of the year's nectar flow, weather permitting, to further boost their winter stores.

A full honey super can weigh 12kg (26lb), so they are unwieldy and awkward to move, especially when you are trying to leave behind as many bees as possible. On a hot day the smell of honey moving around in scorching weather can invite quite the crowd of bees from other hives, and chancing wasps, all looking to rob honey.

Once the super has been safely brought down to the kitchen or bee shed with as few hitch-hikers as possible, it is time to extract the honey. Each honey frame in the super is bulging and golden, with a fresh layer of beeswax cappings to keep in the honey. A large knife, sometimes heated, is required to slice them open, careful to remove as little of the honey as possible, while opening up each hexagonal cell. The sticky beeswax cappings are saved, to be later strained, washed and melted down. The uncapped frame of honey is placed inside the extractor, aka a large metal salad spinner, which usually fits between 4 and 12 frames, depending on size. As honey is thick and viscous, it requires force to remove it from the frames; if you were to hold an uncapped frame of honey face down, only a few drops would drizzle out; most would stay stuck inside the wax comb. The centrifugal force of the spinning drum pushes the honey through the cells and out onto the sides of the drum, to slowly ooze down and collect in the bottom.

From here, the honey is then passed through a sieve or two to remove any remaining wax, bee body parts(!) and any other beehive debris, and is finally ready for bottling. The sticky empty honey frames are returned to the hives, where they are cleverly cleaned by the bees. If this is done early enough in the season, the bees will fill the wax comb with more honey, ready for another harvest or to feed the bees over winter.

For the bees, September is the time that the last of the male drone bees are sent away. Drones are only produced during queen-mating months of the year (spring through summer), so once this time has passed, they are booted out of the hive. The queen begins to stop laying eggs, and the hive is focused on conserving its resources going into winter, and doesn't need extra mouths to feed.

Winter

November through January

If not the meteorological start of winter, November is certainly the start of beekeeping winter in the UK (or at least it used to be; November 2022 was one of the warmest on record). By this time of year, the bees are usually hunkering down into a warmth-conserving cluster as it is too cold to fly. From now until March the bees will be tucked up inside, rarely venturing out of the hive.

The intense physical closeness between bees and their keeper during summer wanes entirely in winter, replaced by reflection and rest after the honey harvest is finished, as both the beekeeper and the bees retreat to their respective warm abodes to shut out the cold. Our relationship has changed, but throughout this time of separation, the bees are always on my mind. Away from the apiary, winter is a time to measure the successes and failures of the beekeeper's craft the previous season, and to prepare for the coming spring. While the bees vibrate their wings in a tight cluster to keep warm inside their hive, the shed becomes my hive of activity, as I happily lose myself in the process of assembling frame upon frame of fresh beeswax, ready for the bees to get to work on once the weather warms. Cleaning up after the honey harvest, cleaning down well-worn tools and bee boxes. An act of love and nurture, the beekeeper-bee equivalent to spending winter preparing a bedroom for an incoming little one, or crafting new kitchen shelves to support your love's culinary adventures. If I haven't already, now is the time to jar up the honey that has been waiting in harvest storage buckets. Memories of each season's blooms and beekeeping efforts come rolling in as the floral aromas rise to meet my nose. Watching the slow molten gold of last summer pour from the spigot to generously fill each glass jar, my heart fills too, with gratitude for the bees' efforts, and the privilege of keeping them.

Cooking with Honey

Our collective sweet tooth for honey pre-dates our love of sugar by thousands of years. The earliest evidence we have of humans and honey is a cave painting in Spain dating to 8000 BCE, depicting a person atop a ladder harvesting honeycomb from a wild nest. A rare natural source of intense sweetness and carbohydrate, honey has been used in healing and faith rituals for years. However, honey can behave in very different ways to sugar, so requires a different approach in the kitchen. Sugar is… sugar. It comes in a powdered, granule or block form. We can fill a spoon with it, we can melt it, we can dissolve it in water and other liquids. We know how it works – it's even polite enough to leave the measuring spoon clean after we've scooped some for cooking. It tastes the same, unless we opt for unrefined, which is to say it often tastes of nothing, it just makes something sweet. Honey, on the other hand, is many things. It's a liquid, but often a solid; it can be runny enough to spill everywhere, or thick enough to bend a spoon and mock anyone trying to prise it from its jar. It comes in a huge range of flavours, all of its own accord, before you even start messing about with infusing it with other things (page 62). It would be a frankly disastrous decision to use lots of bitter tannin-ed Chestnut honey in a mild panna cotta, and a meek Acacia honey wouldn't get a call back if auditioning for a part in a robust beef curry. Here I explore some important culinary characteristics of honey and outline my thoughts on how to use it wisely.

Crystallisation is not the enemy

Crystallisation is a natural change that you will have noticed happening to your jar or squeezy bottle of honey at home, as the honey goes from pourable, to spoonable, to haphazardly microwaveable when you realise that it's solidified at the back of your cupboard and you're in a rush to use some. In the carefully controlled environment of the bee hive, which sits at a balmy 35°C (95°F) and 50–80% relative humidity thanks to the bees' efforts, honey tends to stay liquid in its wax comb, each hexagonal cell sealed shut. Have you noticed that your unopened runny honey bottle stays clear for weeks and months, yet as soon as you open it, it starts to crystallise within a few weeks? This is because you have broken the air seal, introducing the honey to your home's humidity, the odd toast crumb and natural yeasts in the air, which kickstarts the setting process.

Honey is a 'supersaturated' solution of water and sugar, meaning it contains more sugar (above 70%) than would naturally stay dissolved in water at room temperature; it requires specific conditions, otherwise it's a little fickle. The two main sugars in honey are glucose and fructose, and crystallisation occurs when the glucose molecules separate out from the water, forming crystals. The higher the proportion of glucose to fructose, and glucose to water, the quicker it will crystallise.

These ratios originate naturally in the honey's botanical origin – the plant species foraged on by the bees. Different flower nectars have different proportions. For example, Acacia honey stays naturally runny for a long time, but oilseed rape flowers (*Brassica napus*) contain such a high ratio of glucose that the honey can even start to crystallise in the hive; a challenge for bees and beekeepers! We love to drizzle runny honey over yogurt in the morning, and it's undeniably the easiest to use in the kitchen. It's easier to produce too, as thick, set honey is difficult to move through machinery; two reasons why our supermarket shelves are full of squeezy bottles. But how do they stop it setting?

There are two ways of delaying the crystallisation process to produce reliably runny honey: heating and micro-filtering. However, both processes can also damage the flavour profile of the honey, denaturing beneficial enzymes and removing the tiny pollen grains and natural yeasts that give raw honey its wonderful properties.

Despite their slow and sticky nature, thicker honeys are thankfully still available to buy, jostling for shelf space among the ubiquitous squeezy bottles, often labelled 'creamed' or 'soft set'. Clover is a popular and historic varietal of set honey, which dates to at least the 16th century, when the plants started to be sown by farmers as bee-friendly fodder and silage crop. Clover honey naturally crystallises quite quickly, forming small glucose crystals, which give a smooth, non-grainy texture. When you buy a jar of honey from a beekeeper who hasn't used industrial processing, even before opening it you may notice it crystallises eventually; this is a natural process.

While requiring a little more muscle behind your spoon and a deft waggle to coax the slow-motion drizzle, set honeys are the perfect consistency for spreading on crumpets and toast, and the more traditional type of honey that I grew up with. As opposed to heating and micro-filtering which delay crystallisation, you can also bring on the setting process by 'seeding' the honey; stirring some set honey into runny honey, which speeds up crystallisation.

Just like the hue, aroma and flavour of honey, the speed at which it starts to change from runny to set is unique to the honey's botanical origin. It is a metamorphosis to be embraced and worked with, as a characteristic of a live, natural food, like milk turning to yogurt, tea turning to kombucha, or bletting medlars.

The truth about heat

As highlighted above, in the processing of supermarket honey, heat changes the taste and beneficial properties of it. Useful vitamins, minerals and enzymes are denatured at high cooking temperatures and caramelisation occurs, which dramatically changes the flavour. If you are in a supermarket and would like to buy honey that hasn't undergone these changes, it is wise to look for phrases like 'cold-extracted' or 'raw', and ultimately contact the supplier to be sure. Raw honey has a noticeably heady fragrance, lending a deliciously complex floral note to cocktails and delicate bakes.

Fortunately, honey's unique flavour still comes through in cooking, despite the effects of heat, if you are careful about it. Choosing a strong flavoured varietal helps if you are subjecting it to prolonged hot temperatures; a full-bodied Heather or Buckwheat honey will hold its own against the ravages of the oven, while a very mild, delicate honey may get lost along the way. Where possible, add honey at the last minute to preserve as much taste as you can. If some of my recipes seem like I've added a gratuitous drizzle close to the end of cooking, it's intentional, I promise. Lose the flavour of honey by cooking it off and you might as well have used sugar... and then I'd have no book to write!

Honeyed food will also brown more quickly than if you were to use sugar, because fructose caramelises at just 110°C (230°F), while sucrose, the sugar in sugar, needs at least 160°C (320°F) to brown. This means that if you are baking with honey, you need to keep an eye on oven temperatures and adjust accordingly. Hattie Ellis, in her book *Spoonfuls of Honey*, recommends only substituting up to half the weight of sugar with honey, and to reduce temperatures by 25°C (77°F) to counter the faster browning. She also advises on the sweetness of honey: fructose is much sweeter than sucrose, so you need less of it. In my experience, the sweetness of honey can sometimes disappear in baking compared to sugar, and I often find it to be less sweet, as it is also acidic, so my recommended proportions here tend to be a little more generous.

Honey also, quite obviously, has a higher moisture content than sugar, after all it is a 'supersaturated solution' of water and sugar. This makes it a wet ingredient when baking, so you may need to reduce the quantities of other liquids in the recipe if you are adding honey in place of sugar.

Moisture matters / a meringue rant

The higher moisture content of honey relative to sugar is also to blame for my most infuriating yet also most fascinating of recipe test adventures for this book, a honeyed meringue, which I will happily admit here to illustrate the limits of cooking with honey. I love meringue, and so I *had* to include a recipe for one within these pages, however I was not at all satisfied with just drizzling honey over meringue, or roasting fruits in honey and arranging on the meringue. I wanted honey *in* the meringue. Meringue, after all, is just egg white and sugar, so there's plenty of sugar to tinker with. Or so I blindly thought. Try as I might, the results were deeply disturbing in both scent and texture, creating a new biodegradable alternative to florist's oasis and not smelling remotely of honey or frankly anything approaching edible, let alone inviting. Shoehorning honey into meringue became a stubborn mission that I refused to fail at. I was a dog with a bone. I even bought the cheapest dehydrator I could find, having spotted an extremely dubious article on Google extolling the ease with which you could dehydrate honey into powder, which I hoped would be the answer to my meringue physics conundrum. However, having (anxiously) had my dehydrator running through the night for 72 hours straight, creating a sticky mess inside and out, I'm afraid I threw the towel in for a few months, wrote 50 other recipes, then came back to it, and as you can see from the recipe, discovered 1 scant tablespoon to 360g (12½oz) sugar works wonderfully. Happily, it's delicious and very much worth the misguided bother. As I knew deep down all along, quality, not quantity, is key whenever you cook with honey. A little beautifully heady Lavender honey can make a dessert, while too much can ruin it, or at least create a foamy stand with which to arrange your lavender on...

Drinking Honey

My meringue mission often had me crying out for a stiff drink, and luckily honey really is an excellent ingredient for making truly wonderful drinks. Many cocktail recipes require a simple syrup, which is a 1:1 ratio of sugar diluted in water, to better blend with the other liquid ingredients and avoid crunchy sugar grains at the bottom of your drink. But why use boring sugar when you can use floral honey? Also, unlike meringues, the natural liquidity of honey is very handy for such alchemy. Depending on the thickness of your honey of choice, I would recommend gently warming it, adding a tiny splash of water from the kettle, and/or diluting in a 2:1 ratio with warm water to use as you wish, a technique favoured by Richard Godwin in my well-thumbed and splattered copy of his classic cocktailing guide *The Spirits*. A good raw honey can be heady and floral, or deep and rich, so it brings unique flavour even when added in tiny amounts. It is particularly brilliant with elderflower, gin or Bourbon. In this book I've paired smoked honey with mezcal and lime, Buckwheat honey with spiced rum, and if you flip to the frozen section you will discover I have snuck booze into most of the recipes… Eucalyptus honey with vodka, or Campari with watermelon and Acacia honey for two very refreshing granitas, or perhaps Amaretto and honey roast fig ice cream takes your fancy. Both alcohol and honey hinder the formation of large ice crystals, making no-churn ice cream soft and ices delicate. The possibilities are endless!

Fermenting honey

One of my favourite cocktails in this book is a tepache Bourbon sour, the honey being in the tepache, the recipe for which is on page 231. Tepache is a fermented Mexican drink traditionally using the skins of pineapples and their natural yeasts to make something that tastes a little like a fruity kombucha (but that's not doing it any justice at all). Make it immediately; it's your gateway drug into sweet ferments.

If you're anything like me, at the other end of the spectrum of enticing addictive substances is perhaps mead, through no fault of its own. In Europe the mention of mead is likely to conjure thoughts of slightly twee medieval re-enactment groups making 'historically accurate' concoctions that are frankly just the wrong side of rustic to escape novelty or earn a place on my drinks shelf. From at least the Middle Ages in Europe, monks were very often beekeepers. Most monasteries were required to be self-sufficient, right down to their altar candles, so they had apiaries, to produce beeswax (being a much cleaner burn than the cheaper but acrid animal tallow commonly used) and the honey was harvested to make mead, by mixing it with water and leaving to ferment into an alcoholic drink, especially in regions where grapes couldn't be grown. Monk mead production declined as regulation and taxation on alcoholic drinks was introduced, and for many years mead available to buy in Europe was a sort of low-quality monk cosplay offering.

Now, however, in the same manner as cider, there are many brilliant, great-quality, beautiful European meads available to buy without even a whiff of fusty hessian. I highly encourage getting carried away ordering a few, having a little Sunday tasting session in the sun, and making some of your own.

Socks, sandals and Eurocentrism aside, booze fermented from honey and water is considered to be the oldest alcoholic drink known to humankind. Mead was made for thousands of years before Christ and the European monks got round to it, enjoying a much wider popularity in many cultures, from Africa through to China and India. Tej, the national drink of Ethiopia, is a type of mead, infused with the medicinal gesho (*Rhamnus prinoides*), and has been made since at least the 1st century BCE. Ethiopia is the largest honey-producing country in Africa, and 80% of its honey goes to making Tej. Indian mead and honey find their first mention in the *Rigveda*, a canonical Hindu text dating from *c*.1900 BCE, and the very earliest evidence we currently have of honey being used to make a fermented alcoholic drink is from the Neolithic village of Jiahu in China, where pottery dating to 7000 BCE has been discovered with traces of rice, honey and fruit.

Fermenting *in* honey

Fermenting things in honey, particularly garlic, is ludicrously easy and utterly transformative when cooking. Everyone should know what it is and have it in their cupboard, and as you flick through the savoury recipes in this book you will see that I have gotten thoroughly carried away with its application. I should have titled this book *Garlic fermented Honey*. You'll find the 'method' on page 62, but it's just honey, garlic and a bit of time.

Many fruit and veg can be fermented in honey, most successfully anything with a little moisture content. Honey is hygroscopic, meaning it absorbs moisture from the atmosphere around it, so it draws moisture from what you add, then the increased moisture causes the honey to ferment, along with what's in the jar with it. The honey is infused with the flavour of whatever you are fermenting, but in a – God forgive me for using this dreadful word often bandied around fermented food – 'funkier' way, the more interesting sibling of the original fruit or vegetable's flavour. The fermentation eats up some of the sugar in the honey while making it runnier, creating a thin, quite savoury syrup packed with potential for using in cooking. Scotch bonnets are my second favourite to ferment, after garlic.

Fermentation in the hive

When making honey, bees are actually trying to avoid fermentation. Fermentation 'eats' the sugars in honey, which the bees need to eat as their carbohydrate source. To do this, they fan their wings over the honeycomb, which evaporates moisture from the surface, bringing

the water content down low enough (below 20%) to avoid fermentation. The wax cell of honey is then sealed with a fresh wax capping, to keep moisture out. This is why honey is thick and sticky.

Fermenting honey itself is interesting from a culinary point of view for the very reason that bees avoid it, because it eats the sugars. Once you remove the sugar from white sugar, what are you left with? Nothing. No aroma or flavour. Whereas if you remove the sugar from honey you are still left with its amazing floral fragrance and aromas, vitamins and minerals, and colour. This leads to exciting possibilities in bread making, and brewing all manner of fermented drinks.

Bee bread

While avoiding it in the making of honey, bees do use fermentation to make 'bee bread', food for worker bees and larvae. I should probably provide a little more detail, lest you are imagining bees milling wheat grains in their mandibles and lining up tiny bannetons in each hexagonal cell for proving. I haven't yet figured out where the baking would happen in this cute beehive bakery scenario.

Pollen is an important complex food source for bees, which they collect while pollinating flowers by combing the pollen dust off their bodies (like a cat licks its paws and 'combs' its head, only bees have actual combs on their legs for this purpose) and depositing it in leg baskets, like panniers, to fly back to the hive. The pollen is mixed with saliva and packed into wax cells around the developing brood (baby bees), then sealed with a drop of honey. This mixture then ferments, unlocking the nutritional value of the pollen, increasing the bioavailability of phytonutrients and forming bioactive compounds, giving the 'bee bread' higher levels of essential amino acids, vitamins and bioactive antioxidant polyphenols.

Human bread

Pollen grains also find their way into honey, as well as yeasts and lactic acid bacteria (LAB), from the gut flora of the bees and botanical flora during foraging trips, while more yeast strains and microorganisms are picked up during the honey harvest and packaging process. This may prick your ears if you're into sourdough making, as yeasts and LAB are essential components of any sourdough starter. Most sourdough bakers recommend using organic flour, for the presence of wild yeasts that would have been growing on the wheat sheaf in the field, which combine with the yeasts found in the personal terroir of your kitchen to produce fermented magic. Like a sourdough starter, honey also has an acidic pH, which prevents the growth of harmful microorganisms, and contains plenty of glucose and fructose for sourdough yeasts like *Candida humilis* to feed on.

LAB is a useful microorganism that makes sourdough so nutritionally beneficial, as it helps to break down the grain and make minerals more readily available for our guts to absorb. It is also present in honey, via the honey stomach of the bee and the surface of flowers. A recent study produced last year showed that a novel strain of LAB has been discovered in the gut of worker honeybees which holds promising probiotic potential for human gut health.[1]

Can you see where I'm going here? Sourdough and honey are a match made in heaven. Add a teaspoon of raw honey to your organic flour sourdough starter, and you're introducing more wild yeasts, food for fermentation, and lactic acid bacteria to really get things going. In my sourdough recipe (page 140), I've done exactly that, as well as finishing the crust with a garlic fermented honey spray towards the end of the bake. Do give it a go on a lazy weekend; it's delicious.

---

1   Honeybee Gut: Reservoir of Probiotic Bacteria, https://link.springer.com/chapter/10.1007/978-981-16-0223-8_9

Honey Terroir

In defence of 'terroir'

It might seem pretentious to use the term 'terroir' when describing different flavoured honeys, but the diversity and complexity in flavours, created by the particular botanical species foraged on by the bees, deserves full appreciation as a remarkable natural food with huge range and depth to explore. Yet if you go to the honey section of a major supermarket here in the UK, most of the honeys available will simply say on the label 'A blend of EU and non-EU honey'. In stark contrast, wine has long been celebrated for the site-specific flavours a landscape can give to a grape, just as extra-virgin olive oils, coffees and chocolates taste different from different regions. Where previously the food industry always prioritised blending the raw harvest from a huge network of growers to create as much product as possible with the same flavour, now we as a consumer are rejecting this greedy homogenisation, driving a rediscovery of small batches of unique flavour, re-focusing the global market to celebrate regional difference and heritage. It's time to do the same with honey.

Bees have been living the concept of 'food miles' and 'eating local' long before we thought of such slogans to champion small business and combat the downsides of a global mass-scale food system. As the sun comes up each morning, thousands of worker bees set off from the hive to forage around the local landscape for pollen and nectar, taking only what they can reach on wing steam, usually up to around a 3-mile radius, but the closer the better. Honey is the greatest distillation of the natural world, and no two harvests of single-origin honey are the same; what blooms one year may be delayed by frost the next. Celebrating variability and diversity in flavour, using the best of the season, in the best way to make the most of it; raw local honey is the perfect example of this way of thinking about the food we eat.

If you go into your local supermarket, you may struggle to find specific varietals on the honey shelf. National chains prefer producers who can provide enough stock to supply the country, rather than negotiating multiple supplier contracts to sell local honey in each region's branches. And so big brand honey is usually sourced from beekeepers in multiple countries, filtered, heat-treated and blended until it all tastes the same, to produce enough squeezy bottles to ensure every shop never runs out. Everything unique about honey has been removed, and 90% of the flavour gone. While making honey very cheap, this is also very sad. Make a stand and seek out local honey; if you're in the UK, look up your local Beekeepers Association and ask to buy their honey. Try local grocers and farm shops and ask for local beekeepers. Find out what honey from your area tastes like. If you need something cheap and sweet, use sugar. If you're interested in something much more mercurial and interesting, buy local honey.

Before we delve into the flavour range of raw, single-origin honey varietals, a quick definition of terms:

Single origin

Single origin refers to a product that comes from one geographical place, i.e. one apiary, or group of apiaries in close proximity, sharing the same forage area. Commonly misunderstood to mean single botanical origin, i.e. the bees foraged on one plant species.

Monofloral or single species

The correct term for honey made from nectar foraged from one plant species. This is usually achieved by placing the bees' hives within a monoculture: fields of oilseed rape, field bean or lavender, or on heather heathland where little else blooms at a particular point in time.

Polyfloral

Polyfloral honey contains nectar from several plant species. Common polyfloral honeys include meadow, wildflower, forest and orchard.

Raw

Unheated and minimally filtered, the closest to honey straight out the comb. (See page 26 for impact of heating and filtering on the flavour of honey.)

Honey varietals

There are hundreds of types of honey, also known as varietals, and they are defined variously by the forage species, the type of landscape, the community, the country or even the type of honeybee. This book focuses on honey produced by *Apis mellifera*, the Western honeybee, but there several other species of honeybee around the world, including *Apis cerana* from South Asia, and the stingless *Melipona* bee kept on the Yucatan peninsula in Mexico right from Ancient Mayan times.

Below is a short list of some popular and distinct varietals and their flavour profiles, to pique your interest to search for more.

Acacia / Locust

The ultimate gentle drizzle. Very mild, light in colour, and naturally runny. Made from the nectar of False Acacia, the Black Locust Tree – *Robinia pseudoacacia* – which grows easily in the US and Europe, so this is an easy-to-find honey.

Apple Blossom

Harvested from apiaries in apple orchards, where the bees boost fruit yields by pollination. Delicate, fruity and fresh, pale in colour and usually runny, pairs well with fresh, mild cheeses and desserts.

Bramble

A darker amber compared to Apple Blossom, blackberry blossom (*Rubus fruticosus*) honey is a little fruity, generally mild with a tang at the end and soft set. Excellent with gin, so why not make a bramble cocktail...

Borage

A delicious alternative to acacia or 'lime' honey (see overleaf) as it's naturally very pale, clear and slow to crystallise. Borage honey is very mild, with hints of citrus.

Buckwheat

Not a wheat at all, buckwheat (*Fagopyrum esculentum*) is more closely related to rhubarb and sorrel. Buckwheat honey shares some of the same flavour characteristics that buckwheat flour lends to baking: a malty, slightly nutty aroma. The Russian Empire and France used to be huge producers of buckwheat, but it hates high nitrogen, so buckwheat farming declined in the 19th century with the rise of chemical fertilising. Now enjoying a resurgence as an ancient grain and a good crop for low-input farming. Used to make a mead called Chouchen in Brittany, home of the buckwheat galette. Dark in colour with a slightly bitter tone on top of deep spicy notes, this is a very interesting honey to cook and make cocktails with.

Chestnut

A little goes a long way with this dark, aromatic, tannin-y honey with a bitter aftertaste. Try it in a whisky cocktail and hold the bitters. Mostly made from the sweet chestnut tree (*Castanea sativa*), this honey is particularly historic to Italy.

Clover

A historic honey that peaked in the 18th century as clover played a key role in the intensification of agriculture (see pages 42–43).

Eucalyptus

There are many varieties of *Eucalyptus* tree with their own honey in Australia, with flavour profiles ranging from mild red gum to the stronger iron bark. All contain traces of Eucalyptol, and so have the characteristic astringent mint-like tone. The most common Eucalyptus varietal is blue gum, *Eucalyptus globulus*. Widely purported to have extra health benefits, I like to use it in a vodka granita...

Field Bean

A relatively new honey made from using honeybees to pollinate fields of *Vicia faba*, the same bean breed as broad (fava) beans. The plants have extra-floral nectaries on their stems, so bees can harvest nectar before and after flowering. Field Bean honey is pale and mild in flavour, similar to Acacia honey.

Heather

One of the most characterful of honeys, Heather honey even has its own unique texture and word to describe it; it is thixotropic, meaning it becomes temporarily runnier with agitation (stirring), then returns to its jelly-like state. From *Calluna vulgaris*, the ling heather native to Northern Europe, and with strong beekeeping traditions in Scotland and Germany, heather honey is often reddish in colour and strongly flavoured; floral but woody, sweet but tangy, smoky even. Reported to have just as many health-giving properties as the famous Manuka honey.

Lavender

Deliciously floral, as you might expect, and commonly made in France, as you would also expect. Light with little bitterness, a brilliant honey for finishing bakes (see the pavlova on page 194).

Linden / Basswood / 'Lime'

Pale, fresh and even slightly minty, honey from *Tilia* trees is often available as a local honey in cities, where they are popular street trees. Although called lime, this tree is not botanically related to citrus whatsoever, so purposed citrus aromas are wishful thinking.

Manuka

The fashionable superfood of honeys, Manuka tastes reassuringly medicinal and isn't one for your morning toast. The flavour comes from its botanical origin; manuka is a type of tea tree bush (*Leptospermum scoparium*). A highly lucrative export from Australia and New Zealand, manuka can go for £35 a pot. Quite the glow up for a plant considered by the New Zealand government as an invasive scrub weed in the 1950s.

Oak

Predictably deep and woody, but not as bitter or pungent as Chestnut, Manuka or Eucalyptus, with sweet caramel notes, so a little more versatile. Pair it with toasty nutty things to bring out the flavour. Found in many countries with oak woodlands – Greece, Bulgaria and Spain – where it is produced in the same oak groves that feed Black Iberian pigs their much-loved acorns.

Orange Blossom

A beautiful honey with a delicate citrus taste. Often more accurately citrus blossom honey than purely orange, this honey has a tiny amount of caffeine (from the flower nectar) but it's much less than decaf so don't get too excited. Spain has a PDO for Orange Blossom Honey from Granada.

Pine

Pine honey hails from Greece, where it forms 65% of their honey production. It's resinous and herbal as you might imagine, but it's also deep in colour and a little spicy. Formed from honeydew, not flower nectar, which the bees harvest from the rears of tree sap sucking insects. Mmm!

Safflower

An oil crop common in the US, *Carthamus tinctorius* produces a deep reddish treacly honey that I find delicious, and easy to use as it's quite runny. A Californian friend brought me back a jar from a Berkeley farmers' market and I've had to ration it carefully.

Sea Lavender

A rare seaside honey from the UK's East Anglia region, Sea Lavender (*Limonium vulgare*), not at all related to lavender, grows on coastal habitats created by thousands of years of human water management among the mudflats and salt marshes. The honey it produces is pale in colour, mild, with caramel undertones.

Thyme

Gently woody and herbal with a mild, pleasant flavour. Historic to Greece and the Hymettus mountain range, wild thyme honey is produced all around the warm Mediterranean from *Thymus capitata*, *Thymus vulgaris* (our garden thyme) and *Thymus serpyllum*.

Tupelo

Specific to Florida and Georgia in the US, the native range of its namesake tree, this fruity, floral honey is made from the nectar of Tupelo, *Nyssa ogeche*, which grows along rivers and swamps.

Yucatan

Many beekeepers in Mexico now keep the European honey bee *Apis mellifera*. However, Yucatan honey is traditionally made by *Melipona beecheii*, a tropical stingless bee known as Xunan-cab (Regal Lady Bee) by the Ancient Mayans. *Melipona* bees don't make honeycomb, instead storing honey in round wax pots, which can be syringed or crushed to harvest the honey. Tropical honey like Yucatan honey is runnier with a higher water content, slightly less sweet, and has a higher acidity, tasting quite fruity. Yucatan honey is a regional, rather than monofloral, varietal.

Wildflower

A broad catch-all term that doesn't necessarily mean wild in the truest sense, wildflower honey is a multifloral honey that is generally fairly mild, very sweet and brightly floral. A great multi-purpose finishing honey.

Changing terroir – the myth of the idyllic countryside

*'Wild blossoms of the moorland,
ye are very dear to me;*

*Ye lure my dreaming memory
as clover does the bee;*

*Ye bring back all my childhood loved,
when freedom, joy, and health,*

*Had never thought of wearing
chains to fetter fame and wealth.'*

Eliza Cook, 1830

Like humans and animals, bees need a varied diet for optimal health. Flower nectar is their carbohydrate, and pollen is their protein, full of amino acids. However, no one plant species produces pollen with all the amino acids that a bee needs, and so bees must pic 'n' mix a good variety of plant species to ensure a varied diet. You will be completely forgiven for imagining that honeybees thrive best in the most verdant of remote countryside, among vast rolling fields with barely a house in sight.

With the ongoing mechanisation of agriculture, fields and farms have become much larger, losing many hedgerows of blackthorn, hawthorn and other flowering bushes that once divided them, and in the UK there are significantly fewer (97% less since the 1930s) species-rich hay meadows, as farmers turn to silage and more productive crops that require less labour. As flower meadows have declined, so has their varietal: wildflower honey. While oilseed rape is a good food source for many bee species (if it hasn't been sprayed with neonicotinoids), once it has finished flowering across several fields, local bees may struggle to find more forage within flying distance. These days, urban and suburban landscapes are often much more diverse in bee-friendly flowering plants than the rural landscape, as gardens are filled with a variety of flowers, so honeybees often find more forage here than the countryside. Gardens are incredibly important for honeybees and other pollinators, and there are millions of gardens, so do feel encouraged that doing your bit to garden in a pollinator-friendly way (see page 49), or planting particularly helpful plants, will have an impact.

Apple Blossom honey and Orchard honey are two more traditional varietals that have suffered in the UK. Apiaries were often traditionally found in the corner of fruit orchards to ensure good pollination and 'fruit set', but orchards have also declined by 56% due to urban expansion, the importation of cheaper fruit, and the breeding of sterile fruit-tree cultivars which don't require bees' services. (The ghosts of our orchards can be found in suburban streets which supplanted them, with names like 'Orchard Walk' and 'Cherry Tree Lane' – just like 'Meadowview' and 'Haycroft Road').

The growing of clover, a flowering nitrogen-fixing legume cover and fodder crop, was widespread by the 17th century, improving soil nitrogen levels and therefore harvest yields, and contributing to the English agricultural revolution. Out of this boom came Clover honey, a much-loved, delicious and historic varietal, usually softly set, with a particular nostalgia about it, probably because it's just old enough to be our grandparents' grandparents' favourite. However, the quest for more nitrogen continued, and once inorganic fertilisers were invented, bee-friendly clover was largely forgotten about. Luckily, we have now come full circle, as government bodies are recommending farmers return to nitrogen-fixing legumes to reduce reliance on chemical fertilisers and increase biodiversity.

The use of chemicals surged in the mid to late 20th century, focused on making farming ever more efficient, with higher crop yields. Herbicides killed the flowering weeds that bees and other pollinators rely on, and fertilisers created thicker, lusher grass, which out-competes flowering meadow species, creating a dense undergrowth unfavourable to ground-nesting birds and animals, as well as polluting nearby rivers. New pesticides killed half of the insect ecosystem, including aphids, which bees and ants farm for honeydew, and the bees themselves. Neonicotinoids, introduced to UK farming in 1991, have been shown to reduce the honeybee's immune system, and parasitic varroa mites – the most serious pest to the Western honeybee – were first reported in the UK a year later, creating a perfect storm for honeybee decline. The same goes for the US: the first neonicotinoid patented for commercial use came in in 1985, and varroa was first reported in US honeybees in 1987.

The changes in our landscapes are changing our honeys; a jar of honey is a reflection of where we are now, rooted in what came before.

Restocking Nature's Larder

How pollination works, and how to garden for the bees

One worker bee can carry up to 40mg of nectar in their 'honey stomach' on one foraging trip, which is over three times their body weight. One beehive makes a whopping 136kg (300lb) honey (roughly) each year just to survive on, before making the extra that beekeepers like me harvest. Nectar is 40% sugar to water, but honey is twice as concentrated, at 80% sugar, so 136kg honey equates to 272kg (600lb) nectar, which is almost a third of a tonne.

To my maths, this works out as six million eight hundred thousand 40mg bee stomachs' worth of nectar in a year, all for one beehive to have enough carbs. To put this into perspective, the UK's foraging season runs roughly from April to September, which is 183 days, so that's 37,158 full bee stomachs of flower nectar on average being brought back to the hive every single day. All this work to make something that we spread onto crumpets without thinking. It's time we properly appreciated this golden wonder!

It is impossible to understand honey without understanding the relationship between bees and plants, and I find delving into the secret lives of plants and their pollinators an utterly fascinating way into appreciating the marvels of a raw jar of honey sitting in your kitchen. In order to do our best for bees and the hundreds of other important pollinators in our gardens, a little knowledge of their lives goes a long way. Learning about how pollination and foraging works, the huge work that goes into making honey, how flowers have evolved together with the bees, and how plant breeders have developed them further for our gardens, allows us to use our spades and wallets wisely, to give our garden ecosystems a helping hand.

How honey is made: the story of pollination

The harmonious relationship between a plant and its pollinator is the very essence of life. Many plants require sexual fertilisation to make seeds, and this happens when pollen is transferred from one flower to another. This pollen transfer can happen in a variety of ways, the wind being one – most cereal crops are wind pollinated – and insects another. Pollen also happens to be, or rather has evolved to be, a rich and complex protein source for honeybees, and flowers have evolved to look beautiful in order to signal the presence of this food. We can think of these enticing flowers as akin to giant neon American diner signs at the roadside saying HOT FOOD RIGHT HERE, with a large flashing arrow to encourage you to take the next turn; under a UV light many flowers have hidden bullseye target patterns that only pollinators can see.

Bees also need carbohydrate to survive, so plants produce nectar, a watery sugar solution, in their flowers as a further attraction and pollination reward. The flower's nectaries are usually cleverly positioned, so the worker bee must rootle past the pollen to get to them, thus getting covered in pollen dust in the process. A honeybee's fuzzy body hair is in fact perfectly spaced to fit pollen grains, with the correct electric charge to attract and hold on to them. A bee will groom the pollen from their 'fur' using their hind legs, which have adapted comb-like structures, and squash the pollen into baskets, like bicycle panniers or pantaloons, on their legs. However, they do miss bits, so when the bee visits the next flower of the same plant species, some of this leftover pollen touches the stigma of the flower – the long central spike usually with a sticky top for exactly this purpose – and the flower is pollinated.

Once a carb-craving honeybee has rustled through a flower's petals, passed the pollen and located the flower's nectary, it sucks up the nectar, via its proboscis, into its honey stomach. On its way back to the hive, the bees' hypopharyngeal gland releases enzymes which break the nectar's sugars down for easy digestion, and add plenty of beneficial properties for us and the bees along the way. This also helps the sugar be more readily dissolved in a low moisture solution, which in turn inhibits yeast fermentation, which would eat the sugar carbohydrate that the bees need.

On the bee's return to the hive, she regurgitates the nectar solution into the mouth of a 'house' bee, who evaporates the moisture content down to 50%, the perfect level for bees to eat, by 'tongue stropping' the nectar back and forth. She then hangs droplets of stropped nectar inside beeswax cells ready for eating or storing. In the dry hive environment carefully created by the bees, the moisture content of the nectar continues to lower until it is under 20%, which is low enough to prevent the sugars fermenting. It is now shelf stable. The cell is then capped with a fresh seal of wax and is officially honey.

A floral map: how bees find flowers

There are over 20,000 species of bee in the world, and 270 species of bee in the UK. Bees can be specialists, visiting specific plant species, or generalists, visiting many. Honeybees are generalists, working their way to the nectar and pollen of many different plant species.

Worker bees returning from a successful foraging trip, full of nectar or weighed down with their bright baskets of pollen, tell other bees where the best flowers are by dancing across the wax comb in a very precise way that is full of exacting information. It is truly incredible. There are two dances, the round dance and the waggle dance. The round dance is useful for flowers up to 100 metres (330 feet) away, whose scents are easily detectable from the hive doorstep. A returning bee runs in a small circle across the surface of the comb inside the hive, grabbing the attention of bees who are about to set off on a foraging trip, who pick up scents from the dancing bee of the nearby flowers she has visited. For flowers over 100 metres away, which is too far to sniff from the hive, setting off in the wrong direction would be a big waste of energy. To avoid this, bees use the famous waggle dance, which is a detailed map.

The bee will move in a circle to attract others and share scents as before, but she will also run across her circle, with a waggle at the start of the run. The more vigorous the waggle the better the food source, and the number of runs across the circle she does in every 15 seconds inversely correlates (yes I learned that phrase in GCSE science and have always wanted a reason to use it) to how far away the food source is. But what about the direction? Bees orientate themselves in relation to the sun. When a worker bee performs the waggle dance she is on a vertical face of wax comb inside the dark hive. The sun is above, so for map-dancing purposes, it is at 12 o'clock or zero degrees on her dancing circle. If the forage is 30 degrees to the right of the sun from her position on the comb, her little waggle run will be at 30 degrees on her dance circle. The sun moves through the sky during the day, so the dance alters accordingly. Once the other bees have understood this, she feeds them a little bit of the nectar she brought back so they can match it to the right plant when they get there. Not only this, but while dancing, she gives off pheromones and electric fields to further help get the message across. (Bee-sized mic drop.)

How to garden for pollinators

There are many easy and ingenious ways we can help bees and other pollinators when it comes to how we garden, right from the initial design of our space, to the types of plants we choose, and how we look after them. Here I've put together some popular useful tips, together with personal observations from years spent muddy knee'd in flower beds, watching the insect world around me as I work.

A note on flowers

Many garden flowers, like roses and dahlias, are rather fancier than their wild relatives, with lots of petals filling the whole flower head. These are known as 'double' flowers, although in practice they may have many more than just double the number of petals usually found on the wild species. The cause of this is a recessive genetic mutation, which can occur naturally, and is in fact the oldest documented form of floral mutation, having been first observed over 2,000 years ago. Plant breeders liked the look of these unusual forms, and began to breed from them, creating all the classic double flowers we know today: roses, pom pom dahlias, peonies, carnations and the like. Nowadays, plant breeders can persuade a plant to express this genetic mutation to create new double flowers, rather than having to find naturally occurring ones and breed from them.

Many bee-friendly planting guides will point out that these extra petals make it very difficult for a bee or other pollinator to get to the centre of the flower where the pollen and nectar are, which is very true. However, it's more than that, or rather, worse than that, for the pollinators.

To grow those extra petals, the mutation changes other flower parts, such as pollen-producing stamens, into new petals. This means that not only is the pollen harder to reach, as it's hidden under all those extra petals, but there is less or sometimes no pollen under there at all. This highlights the significant consequences we can have when we breed plants for our gardens or farms before understanding our ecosystems. When we prioritise one characteristic, like showy flowers with so many petals that they last for ages in our vases, we might be unthinkingly changing the natural relationship between that plant and its pollinator, with sad consequences.

Think about your other garden plants; a flower cultivar that has been bred to be unusually late or early flowering may be out of step with the lifecycle of the insects that evolved to benefit from it, going over before a particular hoverfly or bee has hatched. Hopefully the converse also applies; an unusually early flowering variety may add to foraging options at a tricky time of year, and while a plant from another geographical region may be thousands of miles away from its co-evolved pollinator, it may provide a new food source for your garden's native pollinators. I'm not saying shun the classic garden rose or those retro dahlia pom poms, but perhaps mix them up with some open flowering varieties too. If you need an excuse to buy more flowers, this is officially it. The more we learn and discover about the side effects of our green-fingered pursuits, the more we can garden responsibly and in harmony with our wildlife.

Plant trees for bees

When you think of bee-friendly plants, packets of wildflower seed mix spring to mind, as these are often marketed to us in garden centres and plant shops to help the bees. Wildflowers are great; they are easy to sow, provide lots of colour, the seed packets are often cheap and they can even be grown in large pots. They make us feel we're doing our bit to reverse the huge decline in our flower meadows (see page 42), and they're increasingly being used to great effect on roadside verges and around towns by inspired local councils who have wised up to the biodiversity and aesthetic value that a field poppy can bring to a concrete jungle.

However, an area of sown wildflowers doesn't provide much nectar and pollen compared to the space it requires. They're also usually annual plants, meaning they die after flowering, so unless you buy more seed or allow the flower heads to set seed, oftentimes that seed packet creates a couple of months of happy colour in one summer, but really gives you more joy than the bees.

So what should you plant instead? Well, firstly, you might have just the spot and circumstances where a wildflower mix is the best option for you, so don't for a second presume that it's not a great thing to do. However, if you do have the room and resources to plant a hardy perennial, a shrub, or even a tree, then we're really cooking on gas. You may not realise quite how much pollen trees produce, because their flowers are small, often green and easy to miss (catkins are actually clusters of tiny flowers). Unless of course you suffer from hay fever, in which case you will be keenly aware when the air is thick with tree pollen raining down in April and May!

To quote my fellow bee and plant lover Sarah Wyndham Lewis of Bermondsey Street Bees, it takes half a football pitch of wildflowers to produce the same amount of bee forage that one lime tree (*Tilia*, not citrus) will happily dish out. It's astonishing how much food trees can provide for bees, especially in that springtime flush. But I know what you're thinking: not everyone, or even most people, has space for a huge lime tree. There are plenty of beautiful flowering shrubs in all sizes, from lavender up to ceanothus, and many much smaller trees, like crab apple or cherries, that produce masses of flowers for bees and yourselves, and only need a decent pot or small flower bed in which to thrive. Small trees in large pots are especially great if you have a small patio space, balcony or, like me, you rent your property, as it means you can take the plunge and commit to a longer-lasting plant and still easily take it with you when you move.

Garden design and planting

If you are lucky enough to have an outdoor space with permission and resources to landscape and plant it up, this section is for you. Garden design is what I do for a living when I'm not busy in a beehive, pruning roses or making a mess in the kitchen, so I'd be remiss not to share my learnings here. However, I am well aware that many people don't have this luxury, so I've included helpful advice that is scalable even for the smallest of outdoor spaces.

1: Maximise borders, minimise lawns

Lawns are popular. Lawns are useful. Kids like to run around on them, dogs like to mess on them, people like to put 'gazebos' on them, expend lots of electricity or fossil fuels mowing them at antisocial hours, or turn them yellow with paddling pools. It's a time-honoured tradition. As such, housing developers usually opt for a thin, sad flower border around a massive lawn, thinking no one wants anything else, and because filling nice borders with nice plants costs them more money. However, if you have a clean slate, or are willing to make some adjustments, try to increase the size of your planting beds relative to your lawn. I promise you won't miss the extra foot or two of grass to mow or pick up dog poop from, and you will have made room for more beautiful flowering plants, creating joy for bees and yourselves.

2: Use your lawn as a canvas, and paint with early bulbs when it's too cold for football

If you can't spare an inch of turf there are still inspiring ways of making this green desert extra beautiful while creating more forage for bees. Try planting small early flowering bulbs throughout in drifts. They only need a slit in the turf to be poked through, so you won't be making big holes in your lawn. They will flower before you need to mow, and the result will knock your socks off. Ham House in London have done this very successfully on their historic formal lawns, and so can you. Crocus, tiny tulips (*Saxitilis*, *turkestanica*, or *linifolia*) and grape hyacinth (*Muscari*) work well. These early flowering drifts are especially helpful for bumblebees, who emerge very early in the year and are entirely dependent on what's flowering then to survive.

3: All green is not grass

Try alternatives to grass if you would like a 'lawn' effect but are happy to experiment. At Royal Botanic Gardens Kew there is plenty of chamomile growing through the lawns, which if you let it flower is a great boon for pollinators, and obviously provides a great calming fragrance too. Flowering thyme and clover also produce beautiful flowers if you let them. Embrace the idea of a biodiverse lawn, with patches here and there flowering, as you would the daisies (more on those later). Lawn 'weeds' like chamomile and clover also stay green long after grass has yellowed in summer drought.

4: Mind the Gap: create a bloom calendar

Think about flowering times when deciding what to plant. Try creating a calendar of when all your must-haves will bloom, then look for the blank spots. There's always something that can fill a blossom gap, and it's easy to find out when something flowers; nurseries or garden centres will tell you, as will the RHS plant finder online.

5: Plant in species drifts, or not

Grouping the same plant cultivar or species together is a particular planting design technique that happens to make foraging especially efficient for pollinators like bees who display 'floral constancy' – the desire to forage on the same species in one trip. However, if you'd like to arrange your plants as a mix or matrix, with cultivars scattered throughout rather than distinct groups, don't worry. Many plants do naturally distribute like this in wild landscapes, and bees are used to moving around. It depends on the design style you like, but I often like to combine the two in a border design, or a woodland ground layer; creating concentrations in some places that scatter out between others. Best of both worlds!

6: Single and ready to mingle

As discussed previously in my floral anatomy lesson (page 49), single open flowering cultivars are best for pollinators, so try to choose those when selecting your plants.

7: Chemical free

Chemicals are just not good for biodiversity and healthy ecosystems. Unfortunately, it's not just how you garden that matters, it's how they garden at the nurseries and garden centres where you bought your plants. Bumblebee hero and conservation ecologist Dave Goulson performed a study in 2017 that tested plants sold in major supermarkets and garden centres for neonicotinoids and other pesticides. Seventy per cent were contaminated, and half of those had an RHS 'Perfect for Pollinators' label. Pesticides such as neonicotinoids remain in plants for several years, coming out in their pollen and nectar, which bees then eat. To avoid this, ask your garden centres and nurseries about their pest management. Are they chemical free? Seeds are also often treated with chemicals, not just plants in pots. While the EU has banned neonicotinoids for most uses, unfortunately the UK has reversed its ban, so it's a real risk.

8: Reach for the skies

Do you have fences or walls? Probably. Are they already covered in climbing plants? Well done! If not, that's a whole area of your garden that could be covered in beautiful flowers for minimum effort, which is very exciting. Some Clematis (*armandii*, *cirrhosa*) flower very early in the year, with beautiful white bell flowers to brighten up a dark winter day and provide a valuable food source for any chilly pollinators.

9: Ivy is your friend

If you've inherited an old ivy (*Hedera helix*), think before you clear it out. Once an ivy has matured, after about 10 years, it will start to flower, producing those strange globe-like structures between September and November. Many plants don't flower at this time, and bees love ivy. Seeing an ivy wall humming with life on a sunny October afternoon is a joy. If you don't like the extra bulk of the flower heads, simply prune them off after they've finished flowering. Ivy used to get a bad rep as it was thought to damage walls, but an Oxford study has revealed this not to be strictly true.[1] Ivy's aerial roots can get into cracks but doesn't cause them, so if you have wall cracks or mortar missing, fill and fix these before growing ivy (or other climbers, which will mask if damage has worsened!). If the masonry is sound and your pointing is in good condition, you've nothing to worry about; simply keep it away from window frames, gutters and fixings. The study also found that ivy forms a very effective thermal regulation blanket, much like any other climbing plant, protecting walls from frost damage and keeping your house a little warmer in winter, while shielding you from excess heat in summer; plants are a valuable tool for keeping your energy bills down!

10: Soften your hard landscaping

Hard landscaping = patios, decking, paths, steps and the like. Nothing for pollinators here, and usually not brilliant for the environment. Hard surfaces absorb the heat of the sun during the day and release it at night, contributing to the 'heat island effect', as well as requiring drainage planning to avoid flood issues. Plants (including lawns) provide a cooling effect and retain rainwater, reducing run-off. If you're laying a patio, consider including pockets between pavers for unobtrusive ground cover planting like a creeping thyme or erigeron, which also work beautifully planted at the base of risers on garden steps. Several 'famous' gardens have paved the way (pun absolutely intended) and are great sources of inspiration; Matt Reese's work on the terrace at Malverleys in Hampshire shows how a beautifully romantic, naturalistic effect can be achieved with this simple technique of sowing seeds into the cracks to break up large expanses of hard landscaping.

11: Get potty

If you can't add seeds or plants into cracks and gaps on your patio, arrange pots instead. They look wonderful and are brilliant for doorsteps, small balconies or courtyards bereft of soil, plus adding plants to that space will help mitigate the 'heat island effect' I mentioned above. You can rearrange them to keep changing things up, get your daily dose of vitamin D when you pop out to water them in between Zoom calls, and if you choose herbs and let them flower, like rosemary and thyme, then you've got food for you and the bees. I know gardeners who have made whole careers out of planting design for pots. It's a wonderful way to express your creativity and every little helps for pollinators.

---

1 Thermal blanketing by ivy (Hedera helix L.) can protect building stone from damaging frosts, https://ora.ox.ac.uk/objects/uuid:89c76782-bc06-4288-b67a-b63313d97053

Garden maintenance

Once your garden is planted, or if you already have an existing established garden, there are still many things you can do, or stop doing, to ensure you're gardening in a pollinator-friendly way. There are many different styles of garden maintenance, and wildlife-friendly gardening is becoming increasingly more popular, so thankfully there's plenty of information about. Here are my tips, and stories, of how to make the most of your garden for the bees, and save yourself a lot of work in the process.

1: Mowing, a faff

Have a go at maintaining your lawn a little differently; it can really help your wildlife. During the early stages of the COVID-19 pandemic, in March and April 2020, Kew Gardens was closed to the public, and many of Kew's 200 gardeners were furloughed, which meant that the 330 acres of grass couldn't be mowed as they would usually. This led to the famous Syon vista, a half-mile long classical highway of lawn, being left to its own devices for a few weeks. It turned white, as far as the eye could see, as 4 acres' worth of daisies (*Bellis perennis*) were allowed to bloom for the first time, creating a huge pollen and nectar source for Kew's pollinators. The bumblebees must have been ecstatic. If you can, allow lawn weeds such as dandelions and daisies to flower before you mow, as they can provide a great food source for bees in early spring. If you're up to it, try 'no mow May', a self-explanatory initiative by conservation charity Plantlife to give garden wildlife a month of new additional habitat to thrive in, while the mower stays in the shed. After May, try mowing a crisp garden path through the long lawn; you'll be surprised at how beautiful it looks, and might be encouraged to keep it that way for a little longer.

2: Clever weeding

Be wise with weeds – common weeds like dandelions and dead nettle are really good for bees, so if you can, at least allow them to 'go over' (finish flowering) before you weed them out. Weeds can also be good 'host' plants, not just food plants. An adult butterfly will feed on the nectar from one plant species, but their caterpillars will need to eat the leaves of another. Many weed species are host plants for butterflies to lay their eggs on. A weed is simply a plant growing in the 'wrong' place. Try to appreciate how perfectly evolved weeds are to growing in tricky, unloved positions, and the useful role they play for wildlife. On Kew's rock garden, grass has evolved a purple colour, as generations of Kew horticulturists and volunteers miss weeds that blend in most successfully with the stone mulch; survival of the purplest.
It is remarkable how weeds respond to their environments, and adapt to survive and thrive, despite our best efforts. Every plant has a valuable ecological niche.

Be careful when weeding and working on dry exposed patches of soil in your flower beds through spring and summer; small, perfectly round holes are solitary bee burrows, so avoid disturbing them if you can. They're fun to spot, for yourself or with children, so crouch down and 'get your eye in' to see if you can find them. If you stay still and quiet, you may even catch its occupant entering or exiting the burrow.

3: Be brave with the scissors

Chelsea chop to extend flowering. This technique can also be used to keep flowers upright without plant supports, as I learned working on Kew's great Broadwalk borders. If you carefully cut the outer stems of a clump of tall flowering plants, they will buttress the clump's inner stems, while flowering a little later and so prolonging the display, giving more bees more time to gather pollen and nectar from that plant. Genius. Plants that respond particularly well to the Chelsea chop include *Rudbeckia*, *Sedum*, *Aster*, *Helenium*, *Phlox* and *Campanula*.

4: Put the chemicals down

As mentioned regarding buying plants, try not to use chemicals, either pesticides or herbicides, when you look after them. Yes, they can provide quick results and everyone is short on time, but we know enough now about the harmful impact they have on our wildlife, so they are best avoided. Plus, working up a bit of a sweat from a spot of gardening is good for your mind and body, so give hand weeding a go, or pinch out the tops of your broad (fava) beans (the tops are also delicious stir-fried) to discourage the imminent aphid arrival, then squish strays by hand, rather than spraying them. Honeybees and bumblebees love the nectar from broad bean flowers, so best not to put them in harm's way with a coating of pesticide on their meal.

5: Find the beauty in death and decay

Increasingly, gardeners and garden designers are allowing plants to die back naturally rather than cutting them down once they've finished flowering, or when they start to turn dry and brown at the end of the summer or autumn. In fact, many designers now choose plants for how they look in this state, as dried flower heads and grasses look beautiful in a winter frost. This change in approach to 'end of season' gardening is hugely beneficial for wildlife, and doesn't necessarily look as 'messy' as you might think with a little selective pruning of anything broken or fallen over. Ladybirds may overwinter inside seedheads (they seem to love *Phlomis*) and thick tufts of golden dried grass provide valuable cover and habitat for many creatures and insects. In early spring, carefully cut back before the new green growth emerges, leaving just a few weeks with bare soil while bulbs emerge and grasses begin to shoot, and the season begins all over again.

Shopping time: plant recommendations

How you garden and what you grow are very important for supporting pollinators and providing forage, but if you're planning your garden or making some changes, where to start when there are thousands of plants available? These days it is thankfully easy to search for bee-friendly plants here, there and everywhere, and there are already many books specifically written on this. Here are some of my favourites, divided by season, so you can slot them into your flowering calendar.

If you live in the rural countryside near lots of farms, have a look to see what crops are flowering around you in June. Farmers and beekeepers often talk about the 'June gap', which can be hard for honeybees. Hawthorn hedges and acres of oilseed rape have finished blooming, but summer wildflowers haven't opened yet. Luckily for gardeners, there are plenty of beautiful plants that flower in June, whether you're looking for a tree, shrub or herbaceous perennial. So if you do live among the fields and notice a shortage of June blooms about, consider adding some to your garden to give local bees and other pollinators a helping hand.

Late Winter into Early Spring

Clematis cirrhosa or *armandii*, *Eranthis hyemalis* (winter aconite), *Berberis aquifolium* (mahonia, oregon grape), *Galanthus nivalis* (snowdrops), *Crocus*, *Helleborus* (hellebore, Christmas rose), *Primula* (primrose), *Corylus avellana* (hazel), *Fritillaria meleagris* (snakeshead fritillary), *Anemone nemorosa* or *blanda* (wood anemone).

Spring

Osmanthus x burkwoodii (sweet box), *Chaenomeles* (Japanese quince), *Ceanothus* (California lilac), rosemary, willow, apple, cherry, crab apple, *Prunus laurocerasus* (cherry laurel), *Lamium* (dead nettle), *Muscari* (grape hyacinth), *Trifolium repens* (clover), *Allium*, blackberry, *Aesculus hippocastanum* (horse chestnut), *Castanea sativa* (sweet chestnut)

Summer

Catalpa bignonioides (Indian bean tree), *Buddleja* (butterfly bush), *Borago* (borage, star flower), *Echium* (Viper's bugloss), *Centaurea* (cornflower), *Nepeta* (catmint), geranium, lavender, *Erigeron* (Mexican daisy), *Oenothera* (evening primrose, beeblossom), *Monarda* (bee balm), *Agastache* (anise hyssop), *Oreganum* (oregano), *Salvia*, *Erica cinerea* or *Calluna vulgaris* (heather), *Cynara cardunculus* or *scolymus* (cardoon, artichoke), *Echinacea* (coneflower)

Autumn

Hedera helix (ivy), Chelsea chopped *Buddleja* (butterfly bush), *Helenium* (sneezeweed), *Hylotelphium spectabile* (ice plant, stonecrop), *Verbena*, *Rudbeckia* (coneflower), *Erysimum* (wallflower)

Winter

Erysimum (wallflower), *Hebe*, *Hedera helix* (ivy), *Arbutus unedo* (strawberry tree), *Viburnum x bodnantense*, *Berberis aquifolium* (mahonia, oregon grape)

Pollen chart

Each plant has a different colour pollen. Beekeepers use pollen charts like this one to read the combs of stored pollen and help determine which plants their bees have been foraging on.

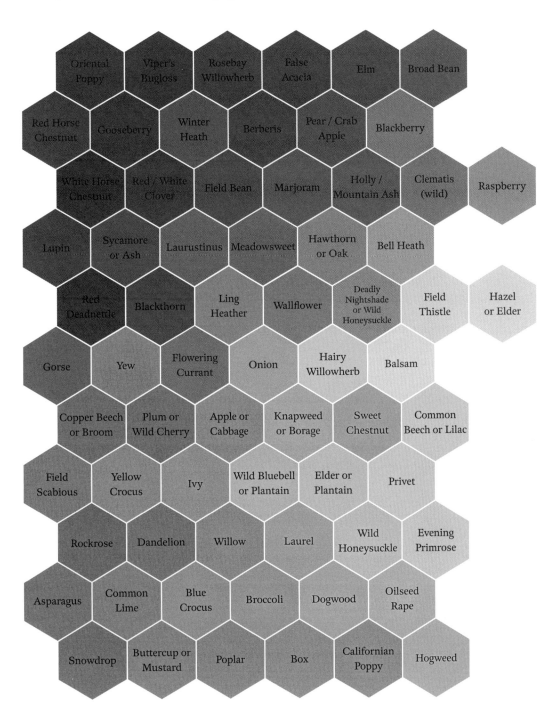

CHAPTER 1

Jars

62	Honey jar ferments
65	Home-smoked honey
66	Honeyed mushrooms, 2 ways
68	Fermented fennel kimchi
72	Harissa buttermilk dressing
73	Sour cherry chipotle hot sauce
74	Lemon curd

Honey jar ferments

RAW HONEY

Honey ferments are gorgeously mercurial concoctions, to be lovingly stockpiled on kitchen shelves, gently tended to over the course of weeks, and sometimes months, as their magical flavours develop.

Honey is hygroscopic, meaning it draws water out of its environment; whether that be from the air, or food submerged in it. This extra moisture starts the process of fermentation (bees intentionally lower the moisture content of honey to avoid fermentation, in order to give it a long shelf life in the hive – see page 31). Both the fermented foods added to the jar and the fermented honey itself can then be used in myriad different recipes. It's an utter joy to experiment with them and discover all the flavour-pairing possibilities.

Ideally the honey should be raw (unheated/minimally filtered) and the ingredients for submerging should be organic and lazily washed (free of dirt but hopefully preserving some natural yeasts on their skins); this is to help the fermentation process. Cutting or bruising the ingredients also helps the honey suck out the moisture. The final fermented honey will become very runny as a result of the increased water content.

Prepare your selected ingredients and add them together with 250g (9oz) honey to a larger jar – around 500g (1lb 2oz). Store at room temperature.

The jar will need inverting and/or stirring every day for the first week, with the jar left unlidded (when the right way up of course) for periods of time to allow microbes from the air to interact with the fermenting honey.

After a week, the turning/stirring can be done every few days, and once the ingredients have sunk to the bottom of the jar it can be left to itself, with the lid on, fermenting away gently. Periodically taste to see how the flavour deepens; some people love to ferment their garlic for over a year!

Suggested ingredients – select one option per jar

1 bulb of garlic, cloves separated and peeled, slits made in the cloves

2 sticks of lemongrass, cut in half lengthways then halved crossways

4 scotch bonnets, halved

2 figs, quartered

1 green mango, peeled and cut into chunks

Handful of gooseberries or sloes

Seeds of ½ pomegranate

¼ quince, peeled, cut into chunks and tossed with 1 tsp water

3 apricots, halved (with stones/pits)

Home-smoked honey

OAK, CHERRY, MAPLE, BIRCH

This is a simple and wonderfully fun way to cold-smoke honey and other earthly delights at home. I encourage any excuse to own or use a beekeeping smoker, even if you have no intention of getting bees; they are such marvellous objects in their own right.

Different woods impart different flavours; try to get hold of a selection to experiment with. Fruitwoods such as apple and cherry are popular, and bolder flavours like oak and hickory will pack a deep punch. The longer you leave the honey in your makeshift smoking chamber, the stronger it will taste. Find your preference for levels of smokiness; have a play and see what tastes best for you.

350g (12oz) jar of good-quality raw honey, or enough to fill a plate that will fit in your smoking chamber of choice

A beekeeper's smoker with insert

Smoking woods of choice in the form of shavings/fine wood dust/chips

A large, tall storage box (clear is more fun)

Pour your honey onto a large, deep plate into a layer around 0.5cm (¼in) thick. This allows for maximum contact area for the smoke to touch. Set the plate on the ground or a table.

Fill the ventilated insert of your smoker with the smoking wood, and set alight. Close the insert and position inside the smoker, closing the smoker lid. By compressing the bellows, gently puff air into the smoking chamber to help the smoker get going.

Once the smoker is steadily puffing away by itself, place next to the plate of honey and cover both with the upturned storage box. Set a timer for 20 minutes, then lift up the rim of the box slightly and taste with a teaspoon to test the smokiness. Keep tasting in 20-minute intervals until you are happy. If the smoke runs out, refuel and light again.

COOKING WITH SMOKED HONEY

Oh the endless possibilities... think of the cocktails: mezcal, Bourbon, peaty whisky, Campari – they all sing with a soupçon of home-smoked honey. Try the margarita on page 238 or the sesame old fashioned on page 241. Smoked honey also works wonderfully well in the meat dishes here in this book, such as gochujang sticky apricot wings on page 99.

Honeyed mushrooms, 2 ways

Both of these honeyed-mushroom situations are easy to make. They add a little something extra to many dishes, such as the parched peas (see page 110) and fennel roast squash (see page 106). Sometimes I like to make both and serve together, a little plump, vinegary bite next to a deep, roasted, half-caramelised one. They're brilliant on toast over a little soft cheese, or added to soups and sauces.

Quick pickled mushrooms

Cut the mushrooms into thick slices, or leave whole if small. Place in a small saucepan with the salt and a 1:1 ratio of white wine vinegar and boiling water – enough to half submerge the mushrooms. Simmer vigorously with the lid on for 10 minutes (with chopped garlic if using), then leave to cool. Once lukewarm, stir through the honey and balsamic vinegar, and leave to pickle for an hour or two.

You can serve these as they are, but I also like to toss them with some chopped fresh herbs and crumbled walnuts. Store in the fridge.

Fills a 350ml (12fl oz) jar

250g (9oz) assorted fresh mushrooms

About ¼ tsp fine sea salt

White wine vinegar

1 tsp garlic fermented honey (see page 62), or 1 tsp honey and 1 garlic clove, roughly chopped

Dash of good balsamic vinegar

Sticky roast mushrooms

Preheat the oven to 140°C/275°F/gas mark 1.

Slice the mushrooms thickly, or leave whole if small, and toss with the olive oil, honey, vinegar, garlic, sea salt flakes and thyme leaves, if using, in a roasting dish.

Roast gently for 10 minutes, keeping an eye on them; they can burn quickly due to the sugar content. Once nicely coloured up and a little caramelised, remove to cool and toss with a little splash of red wine vinegar if you like.

Serves 2, as a topping

250g (9oz) assorted fresh mushrooms

4 tbsp olive oil

1 tbsp honey

1 tbsp good balsamic vinegar

1 garlic clove, minced

Pinch of sea salt flakes

Sprig of thyme, leaves stripped (optional)

Red wine vinegar (optional)

Fermented fennel kimchi

RAW HONEY

Historically, kimchi is made in large earthenware jars, sunk into the earth and left to ferment slowly over a freezing winter. Thus, time and chilling are key to making good kimchi. A short, lively ferment on the worktop, then a slower ferment in the fridge is how I do it, and I always serve it chilled straight from the fridge. Sugar helps aid the fermentation, and brown rice miso is commonly used, but you can also use a little honey. Try to use raw honey if you can for maximum microbial activity.

The key ingredient to find is Korean red chilli flakes, which are relatively mild so the 5 tablespoons or more required for this recipe shouldn't shock you. I've added fennel here for its aromatic edge, which pairs well with the subtle honey note, and of course for its crunch. You can chop up the cabbage wedges for a quicker brine time and an easier effort packing your jars, but I like serving a whole wedge on a side plate, cut crossways, for people to take a section.

Fills a 1-litre (2-pint) jar, or 2 x 500ml (1-pint) jars

1 Chinese (napa) cabbage
2 litres (4½ pints) boiling water
260g (9oz) flaked sea salt
½ plump fennel bulb
½ daikon or handful of radishes, about 120g (4oz)
1–2 carrots, about 160g (5½oz)
3–4 spring onions (scallions)
2 large garlic cloves
5cm (2in) piece of fresh ginger
1 tbsp fish sauce
½ tbsp raw honey
Dash of garlic fermented honey (see page 62)
5 tbsp Korean chilli flakes

Remove any blemished cabbage leaves. Cut the cabbage in half lengthways, and each half into three, creating six segments. Alternatively, you can cut part-way from the base and tear off each segment, for maximum satisfaction.

In a large heatproof bowl big enough to easily fit the cabbage, pour the boiling water over the salt. Stir to dissolve, then leave to cool. Once tepid, add the cabbage wedges, submerging them thoroughly to ensure no air bubbles, and weigh down with anything to hand – I use a large casserole with a taco press inside! Leave to brine until the thickest white stems of the cabbage easily bend rather than snap, from 2 hours until overnight. The cabbage will begin to release water as it brines, so keep an eye on the bowl in case it overflows; scoop out excess liquid as necessary.

Once the cabbage is bendy, prepare the rest of the veg. Using a mandoline on a relatively chunky setting, if you have one, slice the fennel bulb and daikon or radish into beautiful cross sections. Peel the carrot(s) and slice into matchsticks (I use a swivel julienne peeler). Trim the spring onions and cut in half lengthways and crossways.

Prepare your spice paste. Mince the garlic and ginger (I use a microplane) and mix together in a small bowl with the fish sauce and honeys. Finally, add the chilli flakes and thoroughly stir to a paste.

Remove the cabbage leaves from their brine, reserving the brine, and squeeze them thoroughly to remove any excess water. Pat dry. Wearing disposable or washing-up gloves, slather the spice paste through the cabbage wedges, working it into every layer, being sure not to miss bits. You should have a little leftover in the bowl. If not, just make up a little extra.

Place the cabbage to one side, and stir the rest of the vegetables through the remaining spice paste, thinning with the reserved brine if needed, to coat thoroughly.

Time to fill the jar(s). Alternate spoonfuls of vegetables and cabbage wedges, thoroughly packing and pressing everything down to remove as many air bubbles as possible. The cabbage wedges are easier to handle if rolled into little chunky balls.

Add spoonfuls of the leftover brine as you go, to fill up any air gaps. Once the jar is full, push down as much as possible, then add a little more brine on top to cover. Tear off a small piece of baking parchment, big enough to cover the surface and come over the rim of the jar, then add a small weight or object on top to keep everything submerged under the lid – I use wine corks. Loosely screw on the lid(s) to allow some airflow and leave to ferment at room temperature out of direct sunlight for 2–3 days. It should start to smell like kimchi at this point. Transfer to the fridge to continue fermentation slowly, giving it a taste after a week or two (I like mine 3 weeks in). Enjoy with anything and everything.

Harissa buttermilk dressing

YUCATAN HONEY

This lip-smacking drizzle will turn the most pallid iceberg lettuce into a smoky, zingy delight. It also works well gently tossed through charred French beans or new potatoes. The acidity of the vinegar and buttermilk aptly tenderises meat, so do try it as an overnight marinade on chicken thighs. Look, just try it on everything.

Add all the ingredients to a bowl and hand whisk until smooth and emulsified. Season to taste. If dressing salad leaves, do so at the last minute, to preserve crispness.

Makes about 75ml (2½fl oz)

2 tbsp good-quality olive oil
½ tbsp smoked harissa paste
⅓ tbsp runny honey
1 tbsp buttermilk
1½ tsp apple cider vinegar
½ tsp pomegranate molasses
3 grinds of Sichuan pink pepper
Salt

Sour cherry chipotle hot sauce

HOME-SMOKED HONEY

Pronounced chuh-powt-lay, chipotle chillies are ripe jalapeños that have been smoked and dried. You can buy them whole, pre-ground into a powder, or in a sauce with tomato, vinegar and garlic, called 'chipotle en adobo', which when blitzed with mayonnaise makes the best spicy burger sauce ever (lime pickle blitzed with mayo coming a close second, for chicken burgers). Fruit hot sauces are my favourite; dreamy pairings like mango and habanero make my mouth water. Here I've plumped for the jammy tang of dried sour cherries. This one's quite thick; if you like your hot sauce runny, strain this through muslin (cheesecloth) and add more water or vinegar to taste.

Makes 250ml (8½fl oz) +

50g (2oz) dried chipotle chillies

2–4 garlic cloves, peeled

1 bay leaf

2–3 allspice berries

60g (2oz) dried sour cherries

150ml (5fl oz) water

½ red onion

⅛ tsp ground cumin

1 tbsp tomato purée (paste)

1 tbsp home-smoked honey (see page 65)

50–75ml (2–2½fl oz) red wine vinegar

Char the chillies and garlic cloves in a dry frying pan over a moderately high heat until dark spots appear. Set aside the garlic and place the chillies in a small saucepan, together with the bay leaf, allspice and dried cherries. Pour over the water and gently simmer with the lid on for 30 minutes.

Drain off the water and reserve. Discard the bay leaf and add the remaining soaked ingredients to a blender together with the garlic, red onion, cumin, tomato purée and 75ml (2½fl oz) of the reserved soaking water. Blend until smooth, adding more of the soaking water if it seems a little thick.

Transfer to a pan and add the honey and 50ml (2fl oz) of the vinegar. Simmer very gently for 10 minutes, then taste, adding the remaining vinegar to your preference. The sauce may need loosening with a little water or more vinegar. Remove from the heat to cool.

Once cooled, pour into a small bottle or jar and store in the fridge, where it will keep for a couple of months.

Lemon curd

CLOVER HONEY

Mmmmmm lemon curd. A childhood favourite.

Grate the zest of 3 of the lemons and rub into the sugar. Juice all the lemons; you should have 250–275ml (8½–9fl oz).

Whisk the eggs, egg yolks and sugar together, adding the lemon juice gradually. Heat very gently in a saucepan, stirring, until thickened well; be careful not to overheat or stop stirring, as the egg white may form cooked lumps. If it does, strain. Whisk in the salt and honey. Remove from the heat, leave to cool slightly for a minute or two, and transfer to a food processor or bowl for stick blending.

While blending, gradually add the butter, a cube at a time, to the curd until it becomes pale and smooth with no lumps. Decant into clean jars and leave to set. This will keep in the fridge for 1–2 weeks.

Fills 2 x 300ml (10fl oz) jars

6 unwaxed lemons

175g (6oz) white sugar

3 large eggs, plus 4 egg yolks

Generous pinch of fine sea salt

75g (2½oz) honey

100g (3½oz) unsalted butter, cubed, at room temperature

CHAPTER 2

Small Plates

78	Cardamom oats	91	Radicchio pear salad with spiced fritto misto
81	Crumpets – an adventure	92	Guajillo garlic prawns
82	Ginger roasted carrots with chilli and chives	95	Silky squash pasta (or soup)
		96	Broken beans
85	Black garlic and lime tomatoes	99	Gochujang apricot sticky wings
86	Griddled green mustard salad with smoked almonds	100	Achiote orange tacos with jackfruit and cauliflower
88	Charred radicchio and pickled carrot salad with sesame-seared tuna	103	Honeyed chipotle lamb tacos

Cardamom oats

CLOVER HONEY

The night before you'd like to eat these indulgent oats, gently warm the milk with toasted cardamom and Earl grey rooibos tea just before bedtime. It's a wonderfully relaxing way to signal the end of the day and send you happily off to sleep, dreaming of breakfast to come. For me, the calmness of this nourishing ritual rapidly evaporates the next morning, as I have a basic induction hob and the sort of pans that laugh in the face of best efforts to stir constantly, burning the bottom anyway. Happily, said calmness is restored from the first spoonful of this gloriously fragrant bowl of comfort. The bee pollen brings an especially floral, heady top note that is really quite special. Be(e) a little more generous with the ginger in the colder months; it's a warming delight.

Serves 2

4 brown cardamom pods
500ml (1 pint) milk
1 Earl grey rooibos teabag
200g (7oz) oats
¼ tsp ground cinnamon
⅛ tsp ground ginger (double in winter)
Generous grating of nutmeg

To serve
Dollop of clover honey
Sprinkle of bee pollen
Single (light) cream or milk

Before you go to bed, bash open the cardamom pods with a pestle and mortar and toast in a dry pan until fragrant. Gently heat the milk in a pan, adding the teabag and toasted cardamom, and quietly brew below a simmer for 5 minutes. Remove from the heat to cool, then pour into a jar and pop in the fridge. In another jar, add the oats and the same volume of water, shake vigorously, and pop in the fridge too.

The next morning, pour most of the excess water from the soaked oats away, then add them to a pan. Pour the infused milk into the pan through a sieve, to catch the cardamom and teabag, which you can discard. Add the cinnamon, ginger and nutmeg, and stir constantly but slowly over a medium heat until creamy and thickened; don't dismay if the bottom burns a little.

Serve in your favourite bowl, with a dollop of honey, sprinkle of bee pollen, and a little jug (pitcher) of cream or milk, ready to make a moat round the edge.

Crumpets – an adventure

CLOVER HONEY

As a small child I loved playing on my own, joyfully yet furiously absorbed in tiny worlds; rustling under hedgerows, peering over murky ponds, inspecting suspicious jars in my grandparents' understairs pantry, squinting into the subaqueous caverns hollowed out by a dollop of golden syrup on porridge. Naturally, crumpets did not escape my inquisition. Have you ever nibbled the surface off a crumpet to reveal the squeezy labyrinth of bubble holes beneath? Mesmerising. Here, I highly recommend creamed (soft-set) clover honey atop a slab of salted butter, for pure teatime perfection. I haven't added honey into the crumpet recipe itself, because frankly it receives enough mellifluous attention in the topping. I do, however, like to add a little wholemeal or spelt flour and cinnamon to the batter for warming nutty toastiness.

Makes 10

200g (7oz) strong white bread flour

50g (2oz) wholemeal (wholewheat) or spelt flour

7g (¼oz) fast-action dried yeast

1 tsp caster (superfine) sugar

Scant ¼ tsp fine sea salt

¼ tsp ground cinnamon

½ tsp bicarbonate of soda (baking soda)

75ml (3fl oz) warm water

300ml (10fl oz) milk, warmed

Butter, for greasing, plus extra for topping the crumpets

Creamed clover honey, for spreading

Sift the flours into a medium bowl and add the remaining dry ingredients. Whisk to distribute, then add the warm water and milk and whisk into a smooth batter. Cover the bowl with cling film (plastic wrap) and set aside at room temperature for an hour or so until bubbly, frothy and increased in size.

Melt a small knob of butter in a frying pan big enough to hold 4 crumpet rings, over a low–medium heat. Brush the insides of the crumpet rings with the melted butter and add to the pan. After a couple of minutes, use a large spoon to fill each of the rings halfway with the bubbly crumpet batter. Cook over a medium heat until the surface dries out, about 10–15 minutes. Using tongs, flip the crumpets to brown on top for a few minutes, then pop them out of their rings and onto a wire rack to cool. Re-grease the rings with butter to cook the remaining batter.

Slather with butter and honey while still warm.

Ginger roasted carrots with chilli and chives

GARLIC FERMENTED HONEY

Honey roasted carrots (and parsnips, and perhaps all root vegetables) are wonderful. Here, I've gently tinkered with the usual flavours, in favour of ginger and allium, in the shape of garlic fermented honey, nigella seeds and chives. I always keep big jars of homemade or store-bought ginger and garlic pastes in the fridge, for ease when cooking. Do take note of the oven temperatures here and the order of adding ingredients, as ginger and garlic can burn quickly at regular roasting temperatures.

Preheat the oven to 160°C/320°F/gas mark 2.

Place the carrots in a small roasting tray and dot with the butter and salt. Roast in the oven for 25 minutes, tossing halfway through.

Take out the carrots and turn the oven down to 120°C/250°F/gas mark ½. Add the garlic and ginger pastes, nigella seeds, honey and chilli flakes, if using, to the carrots. Toss together thoroughly, then return to the oven for 10 minutes.

Serve the carrots in their tray, sprinkled with a few extra nigella seeds and the chopped chives.

Serves 2

200g (7oz) baby carrots, or regular carrots cut in half or quarters lengthways

Big knob of butter (50g/2oz)

¼ tsp flaky or smoked salt

¾ tbsp garlic paste

1 tbsp ginger paste

1½ tsp nigella seeds, plus extra to finish

1 tsp garlic fermented honey

¼ tsp chilli flakes (optional)

Finely chopped chives, to finish

Black garlic and lime tomatoes

BORAGE HONEY

This salad is sweet and bright, but with deep umami flavours from the tomatoes and black garlic paste. Fresh buffalo mozzarella, torn into large chunks, is an excellent addition to the party. Let the tomatoes sit in the dressing and aromatic leaves for a good hour at room temperature before serving.

Roughly chop the tomatoes, discarding the watery seeds as you go.

Add the black garlic paste, olive oil, smoked salt, lime juice, vinegar, honey and lime powder into a bowl. Toast the sesame seeds in a dry pan, and whilst still hot, pour half the seeds into the dressing bowl and set the other half aside. Whisk the sizzling sesame seeds into the dressing until emulsified.

Combine the tomatoes, basil and dressing, tossing to coat, and set aside for an hour to develop the flavours.

Drain the sliced onion and stir through the tomatoes, then arrange on a plate. Tear over the mozzarella, if using, and scatter with the remaining sesame seeds.

Serves 2 as a starter

2 Iberico tomatoes or similar
½ tsp black garlic paste
2 tbsp good olive oil
¼ tsp smoked salt
Juice of ½ lime
Dash of rice vinegar
½ tsp honey
Pinch of black lime powder
1 tsp black sesame seeds
Small handful of basil leaves, torn
1 sweet white onion, finely sliced, soaked in cold water
1 large mozzarella ball (optional)

Griddled green mustard salad with smoked almonds

GARLIC FERMENTED HONEY

This salad is immensely satisfying. It has crunch, salty smokiness, sweet nuttiness, tangy mustard and a fresh zing from your acid of choice. While I've called this dressing a vinaigrette, I'm not a purist; use lemon juice instead of apple cider vinegar if you prefer more of a citrus acidity. Likewise, I like the punch of Dijon, but wholegrain mustard is equally delicious. If you can't find smoked almonds, and even if you can, have a go at smoking them yourself (see below), or substitute the salt in the vinaigrette for smoked salt.

Put the garlic, honey, vinegar or lemon juice and olive oil into a small bowl and whisk to combine. Add the mustard to taste, whisking thoroughly, then season with salt and pepper.

Heat a griddle pan or dry cast-iron pan over a very high heat. The pan must be hot enough to colour the cos lettuce quickly without wilting. Place the cabbage and lettuce wedges on the pan, leave for 30 seconds or so to char, then turn to colour the other cut side. Remove from the pan. Add the French beans to the pan and shake occasionally until charred, blistered, and cooked through with a bit of bite.

Plate up the cabbage and lettuce wedges and French beans in an unruly pile, and pour over the vinaigrette. Top with the chopped almonds to finish.

HOME-SMOKED ALMONDS

Scatter skin-on almonds over a heatproof tray or plate small enough to fit inside your BBQ (grill). Light a handful of charcoal in a covered BBQ with the vents slightly open. Once hot, move the coals to one side and add a chunk of smoking wood or handful of soaked chips; cherry, oak or hickory would be nice. Cover with the grill rack and place the tray of almonds on the BBQ on the opposite side from the heat. Cover with the lid, vent slightly open, and smoke to taste for around 20 minutes; nibble a nut halfway through to check depth of flavour.

Serves 2

½ hispi cabbage, cut into wedges

2 cos lettuces, cut into wedges

200g (7oz) French beans

Handful of smoked almonds, skin on, roughly chopped

For the vinaigrette

½ garlic clove, minced

½ tbsp honey (smoked or garlic fermented is nice)

1 tbsp apple cider vinegar or lemon juice

1 tbsp extra virgin olive oil

¾ tbsp Dijon mustard, or to taste

Flaked sea salt and cracked black pepper

Charred radicchio and pickled carrot salad with sesame-seared tuna

ACACIA HONEY

This colourful salad combines so many bright flavours and textures that it's just the sort of plate I want to eat at any time of year. Healthy but not worthily so.

First marinate the tuna. Combine the marinade ingredients, whisking with a fork until the honey dissolves; choose a relatively small container, so that all sides are covered once you add the tuna (I use the clear shallow boxes saved from takeaways, as they stack nicely in the fridge). Marinate in the fridge the night before, or morning of, if you can, for maximum flavour.

Next, pickle the carrots. Clean and peel the carrots, then thinly slice lengthways. I like to do this with a knife, usually injuring myself in the process, because I find the slices from a swivel vegetable peeler too thin; I like a little more crunch. Choose a pickling container and combine the vinegar, water, honey and lemongrass, stirring until the honey dissolves, then add the carrots to the liquid and refrigerate for the same length of time as the tuna.

Fifteen minutes before you are ready to eat, prepare the salad. Char the radicchio half in a hot, dry frying pan, then tear off the leaves into a capacious bowl. Drain the pickled carrots and add to the bowl. Combine the salad dressing ingredients thoroughly and pour over the carrots and radicchio leaves, gently turning to coat.

Fill a small, shallow plate with the sesame seeds. Remove the tuna steaks from the marinade, waiting for them to stop dripping, then transfer to the sesame plate and coat the narrow edges in seeds.

In a medium-hot, dry frying pan, sear the coated tuna steaks for about 1 minute on each side, but this will depend on your pan temperature and thickness of the steaks. I like mine raw in the middle and seared on the outside. Remove from the pan and slice thinly (0.5cm/¼in).

Arrange on a shallow bowl and pour over some of the leftover marinade. Plate the salad, sprinkle over some remaining sesame seeds, and serve the tuna alongside.

Serves 2

2 good-quality fresh tuna steaks
½ radicchio
White and black sesame seeds, to coat the tuna

For the marinade
1½ tbsp acacia honey
1 tbsp mirin
1 tsp rice vinegar
4 tbsp light soy sauce
½ stick of lemongrass, cut in half lengthways
Dash of sesame oil

For the pickled carrots
Handful of purple, yellow and orange carrots
100ml (3½fl oz) rice vinegar
100ml (3½fl oz) water
1 tsp honey
½ stick of lemongrass, cut in half lengthways

For the salad dressing
1 tbsp honey
1½ tbsp rice vinegar
Juice of ½ lime
1 tsp toasted sesame oil
Sprinkle of sea salt
Dash of light soy sauce

Radicchio pear salad with spiced fritto misto

WILDFLOWER HONEY

I would merrily dunk the plainest of cardboards into this batter with gay abandon. When you run out of recycling, try unripe pear, fennel and grapefruit, which positively leap out of the fryer shouting for a drizzle of floral honey. I like to fry half and chill half, serving the fried pieces loosely plated among the chilled, with a crisp, bitter leaf – radicchio, the octopus-like tardivo kind if you can get it – for the delicious crunchy contrast of cold raw and hot fried. Tear over some cold mozzarella too if you like, for the sunniest of salads.

Finely slice the pear. Squeeze a drop of lemon into some cold water and toss the pear slices to keep from browning. Set aside a third of the slices for frying and keep the rest chilled.

Cut the grapefruit in half top to bottom. Finely slice one half into half-moons, discarding the ends, and set aside for frying. Cut the peel and pith from the other half, then fillet (supreme) the flesh into skinless segments. Chill the segments with the pear.

Discard any limp outer leaves from the radicchio, cut out the base and separate into firm, crisp leaves. Chill with the pear and grapefruit.

Using a mandoline, finely slice the fennel bulb and set aside for frying.

Gather two small pans. In one, gently warm the honey, then take off the heat. In the other, toast the cumin seeds until fragrant. Transfer the seeds to a pestle and mortar for a light thump and set aside.

Mix the dry batter ingredients with a fork in a large bowl. Next to your hob (stovetop), place a wire cooling rack with kitchen paper on top.

Heat the oil in a deep frying pan over a moderate-high heat. Meanwhile, have your fennel, grapefruit and pear slices ready to one side. Quickly stir the sparkling water/wine/beer/cider into the dry mix with a fork until barely combined. Use a teaspoon to drop a small dollop of batter into the hot oil to see if it's ready; if it starts bubbling immediately and floats to the surface, it's hot enough to fry.

Dredge the fruit and vegetable slices in the batter, then carefully transfer into the hot oil, working in batches of 4–5 at a time according to the size of your pan. Once golden, remove with tongs to the kitchen paper.

Arrange the chilled pear, radicchio and grapefruit segments on a serving plate, and drizzle with light olive oil. Arrange the fritto misto on top, then lightly drizzle with the honey, and some olive oil. Finally, scatter over the crushed cumin seeds and chopped walnuts and serve with a young fresh cheese.

Serves 4

1 firm pear

Squeeze of lemon juice

1 unwaxed grapefruit

3 radicchio, ideally tardivo

1 fennel bulb

100ml (3½fl oz) wildflower honey

1 tsp cumin seeds

800ml–1 litre (1½–2 pints) vegetable or groundnut oil, for frying

Small handful of walnuts, toasted and chopped

For the batter

50g (2oz) rice flour

50g (2oz) cornflour (cornstarch)

½ tsp baking powder

½ tsp salt

Several grinds of cracked black pepper

½ tsp orange baharat (or regular baharat if you can't find orange)

75–100ml (3–3½fl oz) cold sparkling beverage of choice: water, wine, beer, (hard) cider

To finish

Light olive oil, to drizzle

Some young, fresh cheese

Guajillo garlic prawns

HOME-SMOKED HONEY

To some, the suggestion of combining honey, or anything sweet, with seafood is deeply concerning. However, the pairing of salty, fatty and sweet, laced with a kick of smoky chilli and pungent garlic, wrapped around juicy mild prawns, is truly a delight.

You could sub the parsley for wild garlic if it's in season. Bring scallops to the party too if you like, and chorizo is of course a good alternative to pancetta and chilli. If possible, cook the prawns over fire for the wonderful charred bits, but otherwise in a pan or under the grill (broiler) is fine. The prawns cook quickly, so it's best to prepare the other elements first. The honey seems to candy the crispy, salty, fatty pancetta, which is heaven. Have crusty bread to hand to mop up the juices, or serve in a taco with sweetcorn and a tomatillo, lime and coriander (cilantro) salsa, or just by themselves, eaten with cocktail sticks.

Serves 4 as a starter

200g (7oz) shelled and deveined raw king prawns (shrimp)

2–3 tsp guajillo chilli powder, to taste

1 tbsp minced garlic or garlic paste

Pinch of salt

100g (3½oz) good-quality pancetta, diced (or salty bacon lardons, or chorizo)

40g (1½oz) salted butter

1 tbsp olive oil

2 tbsp chopped parsley

Splash of white wine

½ tbsp smoked honey

Toss the prawns in the chilli powder, garlic and salt, then cover and refrigerate until ready to cook.

Add the pancetta to a cold frying pan and turn up to a low-moderate heat. Fry off quite slowly until the fat has rendered and the pancetta has crisped. Fish out the crispy pancetta and set aside.

Turn up the heat under the pan and add the butter and oil, swirling to combine with the pancetta fat until foaming, then add the prawns. After an initial flip-flop to sear on both sides, add the parsley and a splash of the chilled white you are probably already enjoying, while listening to some Marlena Shaw, and cook for 1–2 minutes more on each side. As soon as they turn pink, add the honey and the crispy pancetta, toss quickly, and serve immediately.

Silky squash pasta (or soup)

HOME-SMOKED HONEY

In the early stages of writing recipes for this book, I invited my dear friend and vegetarian, Claire, over for dinner, thinking I'd test out a recipe on her, and promptly realised that all my larger savoury dishes were a tad... meaty. I had a chunk of onion squash spare, some cashew nuts, and the sort of fevered inspiration that only comes from intense pressure. This dish happens to be vegan too, which is fairly unbelievable due to the silky creaminess lent by the silent but hard-working cashews. I originally made this as a pasta sauce, but if you double the amount of veg and thin with a good stock, you have a brilliant soup too, which is delicious topped with the roast mushrooms on page 66. Choose a pasta with plenty of ridges or curls to hold the smooth sauce.

Preheat the oven to 180°C/350°F/gas mark 4.

Mix the paprika and za'atar with enough olive oil to cover the squash, skin and all, and the onion quarters. Place on a baking tray with the garlic cloves and roast for about 20 minutes, until the squash skin and flesh are soft.

Remove and set aside until the garlic is cool enough to handle, then squeeze the flesh from the skins and add to a blender with the squash, onions, honey, nuts, lemon zest and juice, chilli flakes and smoked salt. Add a good slug of olive oil and blend to a fine cream, adding water (or stock) as necessary to achieve a good sauce consistency.

Cook the pasta until al dente, according to the packet instructions, then drain and stir through the sauce.

Serves 2

1 tsp smoked paprika (sweet or hot – your preference)

1 tsp za'atar

Extra virgin olive oil

¼ onion squash, deseeded and sliced into wedges

2 small onions or 1 large, peeled and quartered

2 garlic cloves, unpeeled

1 tsp smoked honey

Handful of cashew nuts

Grated zest of ¼ lemon and a squeeze of juice

½ tsp chilli flakes, or to taste

1 tsp smoked salt

200g (7oz) pasta, such as penne rigate, fusilli or rigatoni

Broken beans

CUPBOARD HONEY

One grey Wednesday in June, I was five short days away from needing to send this manuscript to my editor, and my partner was taking Zoom calls from the bedroom floor having done his back in; we were both thoroughly exhausted. I made these beans, a quick variation on smoky Boston beans, for a speedy lunch. They have everything you need to restore you as you break the back of the day, or your own. Sweet and smoky enough for rich comfort, spicy enough to give you some pep in your step for the rest of the afternoon, and quick enough to feel doable when your mind is full.

I recommend chucking a jar of 'nduja in the shopping basket now and then, to quickly fancify such fridge-raid lunches, but chilli flakes, bacon, pancetta or chorizo would do just fine in this case ('nduja producers, don't come for me; I love you). The toast and egg make it a little breakfasty, which always feels like a rebellious luxury after 12pm, especially if you skipped First Breakfast in favour of a stressed coffee and an urgent to-do list.

Soften the shallot in the oil in a pan over a low heat – I find having a lid on speeds up the softening and reduces the chance of catching. You want it to soften, not fry. Add the garlic after a few minutes and cook gently for another minute or two. Up the heat slightly and add the paprika, fennel seeds, cumin and oregano, with a little extra oil if necessary. Once fragrant, turn the heat down, add 1 tablespoon of the 'nduja and cook for a couple more minutes until softened through. Add the beans, honey, mustard and salt, and a little water to loosen. Stir, then cook over a low heat for 10 minutes or so while you prepare the toast and eggs, adding a little more water if necessary.

Pop the bread in the toaster and poach the eggs. Generously butter the toast, of course, and add a scrape of Marmite, if you like.

In a small frying pan, add the remaining ½ tablespoon of 'nduja, the extra pinch of fennel seeds and drizzle of honey, and warm gently.

Stir the vinegar through the beans, before serving piled on toast, with a poached egg on top, and a little of the fennel 'nduja oil atop the egg to finish.

Serves 2, as a working-from-home lunch

1 shallot, finely chopped

1–2 tbsp vegetable oil

½ tsp garlic paste or 1 garlic clove, minced

½ tsp sweet smoked paprika

¼ tsp fennel seeds, plus an extra pinch

¼ tsp ground cumin

¼ tsp dried oregano

1½ tbsp 'nduja

1 x 400g (14oz) can of baked beans

½ tsp honey, plus an extra drizzle

¼ tsp English mustard

Pinch of salt

Dash of vinegar

To serve

2 slices of bread

2 eggs

Butter, for spreading, and Marmite (optional)

Gochujang apricot sticky wings

HOME-SMOKED HONEY

If the name 'Honey & Spice' wasn't already above the door of London's loveliest food duo's grocery store, it might well have been the title of this book, such is my devotion to spicy food and the magical culinary intersection between heat and honey. The spicy, salty gochujang paste perfectly mellows with smoked honey and tart, sticky apricot to make a joyfully messy, lip-smacking snack, which is just as good with crunchy roasted cauliflower pieces in place of the chicken. You could even make the tepache on page 231 with apricots instead of peaches, and then save the leftover apricot halves to use here. If possible, prepare this 6–8 hours in advance to allow time to marinate. But spicy chicken wings are often an all-consuming sudden craving that can't be delayed, and this is still delicious made off the cuff.

Serves 4

8 fresh apricots

3–4 tbsp gochujang paste, to taste

2 tsp rice vinegar

3 tbsp smoked honey

2 tbsp toasted sesame oil

Splash of dark soy sauce

1 bunch of spring onions (scallions), finely sliced

1kg (2lb 3oz) chicken wings

Sesame seeds, to garnish

Chop the apricots, removing the stones, and toss into a saucepan with a splash of water to loosen up. Simmer, covered, over a low heat until well softened (10–20 minutes), stirring every so often to avoid sticking.

Using a potato masher or wooden spoon, crush the softened fruit to a pulp. Add the gochujang and rice vinegar to the pan, stirring through. Simmer uncovered for 5–10 minutes, stirring and crushing as you go until thickened with no big lumps.

Spoon in the honey and sesame oil and stir vigorously to combine, adding the splash of soy sauce. Remove from the heat.

Stir the spring onions through the marinade, then toss the chicken wings in the marinade until evenly coated, cover and refrigerate for 6–8 hours (or don't).

When ready to cook, preheat the oven to 230°C/445°F/gas mark 8.

Spread the chicken wings onto a baking tray or ovenproof dish. Pop into the hot oven and cook for 30 minutes, keeping an eye on them and basting frequently to keep the glaze on the chicken.

Once the wings are cooked through and have achieved a deep glowing stickiness, remove from the heat, toss and serve immediately with a sprinkle of sesame seeds on top. Goes well with a cold beer or sticky-fingered glass of pet nat.

Achiote orange tacos with jackfruit and cauliflower

YUCATAN HONEY

This flavour combination of earthy, bitter achiote and sharp, sour orange comes from the incredible traditional Mayan dish cochinita pibil, from the Yucatan Peninsula in Mexico. Unsurprisingly, it works beautifully with Yucatan honey, which brings a welcome sweetness, floral acidity and caramelised stickiness. Here cauliflower and jackfruit step up to the suckling pig's plate and shine. Cauliflowers vary in size enormously, so you may need to add a second can of jackfruit to ensure a 50/50 balance, or up the quantities of the marinade to ensure you can coat every crevice.

For the marinade, in a pestle and mortar, combine the achiote paste with the vegetable oil to make a loose paste. Grate in the garlic, add the chilli powder, cumin, oregano and sea salt flakes, then pound together and set aside.

Drain the jackfruit chunks and pat dry with kitchen paper. Divide each chunk in two. Heat a 2.5cm (1in) depth of oil in a frying pan, then fry the jackfruit pieces until golden and starting to crisp up. Drain and set aside on kitchen paper, keeping the oil in the pan.

Break up the cauliflower into individual florets the size of button mushrooms. Mix together with the jackfruit in a large mixing bowl and pour over the marinade. Stir to coat evenly, and marinate for 4–8 hours, or as long as you have time.

Preheat the oven to 200°C/400°F/gas mark 6.

Spread the cauliflower and jackfruit out in a single layer on a baking tray or two, ensuring there is enough space to allow the pieces to roast nicely rather than steam together. Drizzle over some of the oil used for frying and roast for 20–30 minutes until nicely browned and crispy.

Meanwhile, prepare your toppings and have them to hand. Turn the oven off while you heat the tortillas in a hot, dry frying pan until fragrant and slightly charred on both sides. Stack between 2 folded tea (dish) towels to stay warm and soft.

Remove the tray/s from the oven and immediately squeeze over the sour citrus juice, then drizzle over the honey. Stir through and serve at once on the warm tortillas, with the other suggested taco toppings.

Serves 2 (4 tacos each)

1 x 400g (14oz) can of jackfruit in brine, drained

Vegetable oil, for frying

1 medium cauliflower

2 tbsp sour citrus juice (lime, grapefruit, Seville orange)

2 tbsp Yucatan honey

For the marinade

2 tbsp achiote paste

2 tbsp vegetable oil

3 garlic cloves

1 tsp chilli powder

1 tsp ground cumin

2 tsp dried Mexican oregano

½ tsp sea salt flakes

To serve

8 small corn tortillas

Good guacamole

Pickled jalapeños

Sour cream

Tajin seasoning

Torn coriander (cilantro) leaves

Honeyed chipotle lamb tacos

TUPELO HONEY

OK, I just love tacos. One of my favourite things about Mexican cooking is that the ingredients work hard in a relaxed, unpretentious way, providing huge flavour, vibrant colour and impressive chemistry. The apple cider vinegar in this marinade serves to tenderise the meat, much like the sour citrus in the vegetarian take on cochinita pibil on page 100. If you can, plan ahead and leave to marinate overnight for the greatest effect; your taste buds will thank you.

Combine the dry marinade ingredients in a pestle and mortar, then add the garlic and pastes and pound, before adding the vinegar to loosen. Rub the marinade over the lamb chunks, thoroughly coating them, and refrigerate, covered, for 8–24 hours.

Preheat the oven to 160°C/320°F/gas mark 2.

Spread the onion slices over the base of a roasting tray small enough to fit the lamb chunks snugly. Place the marinated lamb on top of the onion and tightly cover the tray with foil. Roast for 2 hours.

Stir together the chipotle paste, honey and oil. Remove the lamb from the oven and stir through the chipotle and honey mixture, being careful not to break up the lamb chunks. Return to the oven, uncovered, for about 10–15 minutes, until nicely browned.

While the lamb is browning, heat a dry frying pan over a high heat and warm the tortillas on each side until fragrant and slightly charred. Keep warm and soft inside a couple of tea (dish) towels until ready to serve.

Stir the lamb and onions, breaking up the lamb a little into smaller pieces, and squeeze over the lime. Serve spooned into the tortillas, with the other garnishes piled on top.

PICKLED RADISHES

Top and finely slice a bunch of radishes, ideally using a mandoline. Submerge in equal parts cider vinegar and sugar and marinate in the fridge for 30 minutes–1 hour.

Serves 4 (4 tacos each)

750g (1lb 10z) boned lamb leg, cut into big chunks

1 onion, sliced

2 tsp chipotle paste

2 tbsp tupelo honey

2 tbsp oil

½ lime

For the marinade

1 tsp coriander seeds, toasted

1 tsp cumin seeds, toasted

2 tsp dried Mexican oregano

1 tsp salt

4 garlic cloves, peeled

2½ tbsp chipotle paste

2 tsp achiote paste

100ml (3½fl oz) apple cider vinegar

To serve

16 small corn tortillas

Guacamole or avocado slices

Coriander (cilantro) leaves

Finely chopped white onion

Lightly pickled radishes (see note)

Tajin seasoning

CHAPTER 3

Large Plates

106	Fennel roast squash with burrata, pickled mushrooms and crispy onions	122	Double plum ribs and lime leaf slaw
109	Harissa hazelnut beetroot with raw cavolo and dilly red peppers	126	Smoked lime and honey chicken
		129	Dark beef and aubergine curry
110	Paprika parched peas with butter-roast fennel and garlic green peppercorn mash	130	Goat biryani with jackfruit and lime leaf rice
113	Minted lamb chops with herbed grains, pomegranate and sherry vinegar	132	Spice cupboard goat with preserved lemons
116	Saikyo salmon	136	Honey hong shao rou
119	Smoked duck, smashed cucumber, slapped noodles		

Fennel roast squash with burrata, pickled mushrooms and crispy onions

THYME HONEY

This is a nourishing and satisfying midweek dinner that'll put a smile on your face. It is also decadent enough to impress friends on the weekend while being genuinely easy to pull together. If you don't like the faff of frying onions on a week night, cover the pan with a sieve to stop the splatter, or fry a batch at the weekend; they keep well and are great atop... everything! Do have a go at the quick pickled mushrooms too (see page 66); their earthy, sharp tang is a delight and they keep well in the fridge for a week or two.

Preheat the oven to 180°C/350°F/gas mark 4.

Wash and dry the squash, then slice in half and scoop out the seeds.

Partially crush the cumin and fennel seeds in a pestle and mortar, then mix with the olive oil, smoked salt and honey. Rub this over the skin and cut sides of the squash. Place on a baking tray and roast in the oven for 30 minutes.

While the squash is roasting, heat a depth of 2.5cm (1in) vegetable oil in a frying pan over a moderate heat. Have a wire rack topped with kitchen paper ready.

Add the sliced onion to the oil (you may need to do this in batches) and fry until lightly golden; stop before they turn dark brown. Remove and leave to drain and crisp up on the kitchen paper.

Wash, dry, and remove the thickest parts of the cavolo nero stalks. Cut the leaves into generous slices, then add to a mixing bowl with the lemon juice, extra virgin olive oil and flaked sea salt. Using your hands, rub the dressing into the leaves, until all are shiny and their waxiness has softened. Chill until serving.

Once the squash is done, by which time the flesh is soft enough to insert a butter knife with no resistance, remove from the oven and leave to cool for a minute.

Arrange the cavolo nero on a plate and sprinkle over the hazelnuts. Add the roasted squash, with the burrata on top. If using, spoon some pickled mushrooms over the burrata. Drizzle the plate with a little extra virgin olive oil, season with cracked black pepper and scatter over the fried onions.

Serves 2 as a main

1 small squash, preferably with edible skin, such as butternut, onion or delicata, about 650g (1lb 7oz)

1 tsp cumin seeds

2 tsp fennel seeds

4 tbsp olive oil

1 tsp smoked salt flakes

1 tbsp thyme honey

Vegetable oil, for frying

1 small onion, finely sliced

6 big leaves of cavolo nero

Juice of ½ lemon

2 tbsp good extra virgin olive oil, plus extra to finish

Large pinch of flaked sea salt

Small handful of roasted hazelnuts, finely chopped

1 small burrata, drained

Pickled mushrooms (optional; see page 66)

Cracked black pepper

Harissa hazelnut beetroot with raw cavolo and dilly red peppers

YUCATAN HONEY

A while ago, 'everyone' was massaging raw kale for 'clean-eating' 'green goddess bowls', and it all felt a little faddy for me. However, I love deeply dark and dramatic cavolo nero, so I gave it a perfunctory half-hearted massage, cringing as I went, and was pleasantly surprised. After a light tussle with some olive oil, good flaky salt and lemon juice, the waxy cuticle (yes, botany) of the cavolo nero leaves rapidly gives up the ghost, and the whole thing suddenly becomes a gloriously intense salad with bite, rather than feeling like you're chewing plastic leaves, as you might imagine raw kale would.

Preheat the oven to 180°C/350°F/gas mark 4.

Toss the beetroot with a little sunflower oil and salt, and roast on an oven tray for 25 minutes.

Meanwhile, thoroughly mix the yogurt, garlic, a glug of extra virgin olive oil, a pinch of the chopped dill and salt and pepper to taste. Transfer to the fridge.

Take a sharp knife and cut out the stems of the cavolo nero, then cut the leaves into long diagonals. Place in a small mixing bowl and add a glug of extra virgin olive oil, a sprinkle of salt and a squeeze of lemon juice. Get in there and thoroughly massage the dressing into the leaves for a minute, until they are soft and shiny. Chill until ready to serve.

Toast the chopped hazelnuts in a dry pan for a couple minutes until fragrant, and set aside. Char the peppers over the hob (stovetop) if you're cooking on gas, or in a very hot, dry frying pan if not. Once nicely blackened all over, you can either remove and cover with a plate to continue softening them or leave them uncovered to retain a bit of crunch; up to you. I leave the stems on, but you can remove them together with the seeds at this point, if you like. Either way, once they've cooled enough to handle, toss them with the remaining dill, a pinch of flaked sea salt, a squeeze of lemon juice and a glug of extra virgin olive oil. Set aside.

Mix together the honey, harissa, cinnamon and another glug of extra virgin olive oil. Remove the beetroot from the oven and chop each half into 2 or 3 segments. Quickly toss the segments in the honey/harissa mixture, and return to the oven for a further 10 minutes until sticky and just a touch crispy round the edges.

Arrange the hot beetroot, warm peppers, and chilled cavolo together, and top with the chopped toasted hazelnuts and seasoned yogurt.

Serves 4

500g (1lb 2oz) beetroot (beets), trimmed, scrubbed and halved

A little sunflower oil

250g (9oz) Greek yogurt

½ garlic clove, grated

A few glugs of extra virgin olive oil

Bunch of fresh dill, finely chopped

Handful of cavolo nero

½ lemon

100g (3½oz) hazelnuts, roughly chopped

2 long red (sweet) peppers

2 tbsp Yucatan honey

2 tbsp harissa of choice

½ tsp ground cinnamon

Sea salt and freshly ground black pepper

Paprika parched peas with butter-roast fennel and garlic green peppercorn mash

FIELD BEAN HONEY

This dish is luxurious to a fault. The best kind of nourishing, dollopy dinner on a cold evening. Parched peas are a northern British snack; a type of brown pea called carlin is 'soused' in vinegar before serving, often in a little paper cone. You can use Puy lentils in place of parched peas, but I urge you to order them online and have a go. They're wonderfully large, like chickpeas, and come out of a slow cook still perkily round with a bit of bite.

A swirl of vinegar or squeeze of citrus enlivens any rich stew just before serving, hard-steering the flavours towards vibrant and away from sluggish territory, no matter how long they've been cooking. If you fancy a little meat here, add a little good-quality pancetta to the parched peas, or serve with fennel-seed-crusted pork chops.

Roasting fennel like this is not done enough; soft, aromatic and buttery. I have Gill Meller to thank for the idea of brined green peppercorns in mashed potato, after a gorgeous dinner at Little Duck Picklery, eating dishes from his Time cookbook, where he plates a huge round Cumberland sausage on top of a cheesy peppercorn mash. Serve with a side of wilted, deeply dark greens tossed in a little butter and apple cider vinegar, if you're feeling a notable absence of chlorophyll.

Serves 2, with leftovers

400g (14oz) carlin peas, soaked overnight in water, or Puy (French) lentils

2 carrots, finely diced

Glug of vegetable oil or knob of butter

2 shallots, sliced

2 garlic cloves, crushed

1 tsp chipotle powder

1 tsp paprika

½ tsp celery salt

700ml (1½ pints) water

½ tsp finely chopped fresh chilli

1 tsp field bean honey

Knob of salted butter

Generous lashing of malt vinegar

Chopped parsley, to serve

For the fennel

1 plump fennel bulb

2 generous knobs of salted butter

4 sprigs of thyme

Freshly ground black pepper

1 tbsp garlic fermented honey (see page 62)

Just before the peas' soaking time is up, trim the fennel, setting the bulb aside, slice the stalks and add to a medium-hot frying pan with the carrots and glug of oil or knob of butter. Add the shallots and garlic, stir through and allow to soften for a few minutes. Add the chipotle powder, paprika and celery salt. Stir for 2 minutes over a moderate heat until fragrant, then add the drained peas or the lentils and the measured water. Pop the lid on and simmer gently for 1–2 hours, stirring occasionally, until the peas have softened. Set aside to rest for 10 minutes, with the lid on.

While the peas are simmering, preheat the oven 180°C/350°F/gas mark 4.

Continued

Tear off two pieces of foil, around 30cm (12in) long. Cut the trimmed fennel bulb down the centre into two halves. Place each half on its own piece of foil, cut side up. Add a generous knob of butter, 2 sprigs of thyme and a few grinds of black pepper to each. Wrap up and place in a small ovenproof dish to catch any drips. Slice the top off the bulb of garlic (for the mash), drizzle with scant olive oil over the top, tightly wrap in foil and add to the dish.

Bake for 45 minutes until all are soft. Remove from the oven and set aside the garlic bulb. Increase the oven temperature to 220°C/425°F/gas mark 7. Open up the foil on each of the fennel halves, brush the cut surfaces with the garlic fermented honey and return to the oven until browned.

Next, boil, drain and mash the potatoes, whipping in the honey, Parmesan, butter and peppercorns. Squeeze the roasted garlic cloves into the mashed potato and gently stir through to distribute.

Stir the chopped chilli, honey, butter and malt vinegar through the peas. Serve piled high next to the mash and fennel, with plenty of chopped parsley sprinkled over.

For the mash

1 bulb of garlic

Drizzle of olive oil

500g (1lb 2oz) Maris Piper potatoes, peeled

1 tsp garlic fermented honey (see page 62)

75g (2½oz) Parmesan, grated

100g (3½oz) salted butter

1 tbsp brined green peppercorns (drained)

Minted lamb chops with herbed grains, pomegranate and sherry vinegar

ORANGE BLOSSOM HONEY

I love lamb chops. And by chops, I have learned that I personally am referring to cutlets, which are specifically rib chops (rather than the meatier loin chop) with the nifty bone handle, great for dipping, with extra tender meat. The fat is delicious when properly rendered, and with the right amount of salt, few meats are more addictive. Mint is unquestionably necessary, pomegranate and Middle Eastern spices are sparkling additions, but it's the sherry vinegar and honey (of course) here that really make the flavours sing, cut through by the garlicky yogurt. The herbed grains are a sort of tabbouleh, using giant couscous (maftoul) and bulgur wheat nestled among the red onion, but you could use regular couscous, or chickpeas.

First, marinate the lamb. Combine the rub ingredients and thoroughly coat the lamb chops, working it into the nooks and crannies and around the bone. Refrigerate, ideally for 4–8 hours, but often I only chill for the time it takes me to make the glaze and yogurt and heat up the frying pan.

Prepare the yogurt. Grate the garlic cloves into the yogurt or labneh, add a generous sprinkle of sea salt and several grinds of black pepper. Stir thoroughly, adding the olive oil and lemon juice as you go. Pop in the fridge until ready to serve.

Add a little olive oil to a moderately hot saucepan big enough to hold a litre (2 pints) of water. Add the giant couscous and bulgur. Swirl to coat all the grains in oil and cook for a few minutes until toasty and fragrant. Pour over the measured water and add 1 teaspoon of fine sea salt. Bring to the boil, then simmer for 5–10 minutes, until cooked through but not mushy; they should be al dente. Drain and toss in a mixing bowl with the extra virgin olive oil and pul biber.

Roughly tear the parsley and mint leaves and add to the grains with the pomegranate seeds. Mix through gently with your fingers or a fork.

Cut the red onion half into 2 quarters and thoroughly char the cut sides in a hot, dry pan. Split out the layers into petals and toss through the herbed grains.

Continued

Serves 2

4 lamb chops
Sea salt and freshly ground black pepper

For the rub
½ tsp cumin seeds, toasted
½ tsp dried mint
½ tsp sumac
½ tsp sea salt flakes
¼ tsp cracked black pepper
1 tsp oil

For the yogurt
2 garlic cloves
300g (10½oz) Greek yogurt or labneh
1 tbsp extra virgin olive oil
Squeeze of lemon juice

For the herbed grains
Olive oil, for cooking
100g (3½oz) giant couscous (maftoul)
100g (3½oz) bulgur wheat
650ml (1¼ pints) water
50ml (2fl oz) extra virgin olive oil
Large pinch of pul biber chilli flakes
20g (¾oz) parsley leaves
20g (¾oz) mint leaves
Seeds of ½ pomegranate
½ red onion

Whisk the glaze ingredients thoroughly to emulsify, and set aside.

Heat a frying pan to a moderately high heat, then add the lamb chops, fat side down first (use tongs to hold them in place), to ensure a good rendering. This should take a couple of minutes. Flip to sear the meat on both sides for another minute each. The lamb should still be pink in the middle, so won't take too long. Set aside to rest while you are plating.

Arrange the herbed grains and yogurt on plates, then toss the hot chops in the glaze and arrange on top. Pour over the remaining glaze and serve.

For the glaze

1 tbsp olive oil

1 tbsp pomegranate molasses

1 tbsp orange blossom honey

2 tsp sherry vinegar

Saikyo salmon

SEA LAVENDER HONEY

Saikyoyaki is a Japanese method of marinating fish in sake, mirin and sweet Saikyo miso from Kyoto for several days, giving fatty fish a buttery, melting texture. It is often done with gindara or 'black cod' (actually not in the cod family at all), and these days is a popular dish at expensive London restaurants. However, it's surprisingly easy to make. Mirin is very sweet, so here I have substituted half of the mirin for honey. Salmon is a great, commonly used alternative to black cod, as is Spanish mackerel, or any fatty fish. Saikyo sweet miso has a particular flavour, so if you can't find it and do substitute with white miso, you will need to double the mirin and honey quantities to make up for the missed sweetness.

Serves 2

2 salmon fillets, as fresh as possible, skin on, scaled
3 tbsp sake
4 tbsp Saikyo (sweet) miso
1 tbsp mirin
1 tbsp sea lavender honey
Oil, for greasing
Sea salt flakes

Sprinkle enough salt flakes over the salmon fillets to lightly coat, and leave to one side for 30 minutes to draw out moisture and firm up the fish. Rinse the fillets in 2 tablespoons of the sake to remove the salt.

Mix together the miso, mirin, honey and remaining 1 tablespoon of sake, whisking to combine thoroughly.

Coat all sides of the salmon in the marinade. Select a marinating container small enough to keep the marinade on all sides of the fish – I actually do this using 2 sheets of cling film (plastic wrap) rather than placing the fish in a bowl or tub. Refrigerate for 2–3 days.

Preheat the oven to 200°C/400°F/gas mark 6 and place an ovenproof frying pan over a high heat. Unwrap the fillets, remove some of the excess marinade, lightly oil the pan and place skin side up in the frying pan. The underside will brown and blacken over a few minutes, then flip to repeat skin side down. Transfer to the oven to bake for about another 10 minutes, until the fish is gently cooked through but not tough.

Serve as it is, alongside sticky rice and wilted greens, or even in a taco with a scrape of sour cream, fresh shiso leaf and lightly pickled radish. Everything tastes good in a taco.

Smoked duck, smashed cucumber, slapped noodles

SAFFLOWER HONEY

This comforting, spicy dish has a brilliant depth of flavour and plenty of contrasting elements to keep you entertained, both in the making and the eating. The slapped noodles are 'biang biang' noodles, from X'ian in China, named onomatopoeically, for the sound they make when slapped down onto a work surface during the very fun hand-pulling technique. The traditional Chinese character for these noodles is one of the most complex, with as many as 58 strokes. Luckily they're much simpler to make than to write.

I've paired them here with sesame smashed cucumber salad, one of my favourite dishes served in my local Sichuan restaurant, and fatty smoked duck infused with Sichuan peppercorns, for their salivating citrus kick. If you can find unhulled sesame seeds, use these for a deeper, toastier flavour. You can buy ready-made Chinese sesame paste, but don't substitute it for tahini; the two are very different!

Pat the duck breasts dry and score the skin into a fine criss-cross, using a very sharp knife. Toast the Sichuan peppercorns, sesame and cumin seeds until fragrant in a hot, dry pan, then lightly crush in a pestle and mortar and combine with the remaining rub ingredients. Coat all sides of the duck in the spice mix, place on a plate lined with kitchen paper, cover and refrigerate for at least 30 minutes, but preferably 4 hours.

Meanwhile, make the noodles. Stir together the flour, water and salt in a bowl and, once it comes together enough, turn out and knead with your hands. The dough will get smoother and smoother until satisfyingly soft and springy. At this point, cut the dough like a cake into 8 equal pieces (for 8 massive noodles) using a knife or dough scraper, then knead for a couple of minutes each. Roll them into little cocoon oblongs, coat with oil, and pop on a plate. Place in the fridge, covered, to rest and prove for 1 hour, to develop the gluten. The noodles will be cooked last minute, while the cooked duck is resting.

While the noodles are proving, make the smashed cucumber salad. Cut the cucumber in half lengthways, scoop out the seeds, then gently bash each half with a rolling pin, cut side down, until it splits a few times lengthways but before you've beaten it to a pulp. Cut into 2cm (¾in) slices, then toss with a big pinch of salt and sugar, and leave in a sieve over a bowl or the sink for about 30 minutes to draw out moisture.

Continued

Serves 2

2 duck breasts, skin on
1 tbsp dark safflower honey

For the rub

1½ tsp Sichuan peppercorns
¾ tsp sesame seeds
¾ tsp cumin seeds
1 tsp Korean red chilli flakes
Large pinch of ground ginger
Large pinch each of flaked sea salt and cracked black pepper

For the noodles

250g (9oz) plain (all-purpose) flour
125ml (4fl oz) water
Large pinch of fine sea salt
Oil (sunflower, groundnut, or wok sesame oil), for brushing

For the cucumber

½ cucumber
Big pinch each of salt and sugar
2 tbsp sesame seeds
1 tbsp toasted sesame oil
½ tsp chilli oil
Dash of dark soy sauce
Dash of light soy sauce
½ tsp rice vinegar
¾ tsp honey
1 small garlic clove, grated

Toast the sesame seeds in a hot, dry pan until fragrant, then finely crush to a paste using a pestle and mortar. Add the remaining cucumber ingredients and stir thoroughly. Once the cucumber has drained, quickly rinse off the salt and sugar, pat dry and toss in the sauce. Cover and refrigerate for at least 30 minutes or until you are ready to serve.

Time to smoke the duck. You don't have to do this; you can skip straight to cooking it in a pan, but I've added this step as an extra delicious option. You will need a meat thermometer.

To smoke: Remove the duck breasts from the fridge, pat dry again, and place skin side up on a wire rack above a drip tray. Light a small handful of charcoal in one side of a covered BBQ (grill). Once hot, add a chunk of smoking wood (I favour cherry for duck), place the duck tray on the grill above towards the other side of the BBQ, insert the meat thermometer and cover, with the vents slightly open. Smoke for about 10–12 minutes and remove if/when the temperature gets to 45°C/113°F. Save any fat/juices that may have collected in the drip tray. Remove the duck from the BBQ.

To cook: Warm and thin the honey slightly with a drop of hot water, and have to one side ready with a basting brush. Heat a frying pan to a moderate heat, and place the duck skin side down in the pan, to render the fat. Don't heat too high or the spices will burn. After a couple of minutes, pour off the fat and put to one side, flip the duck and brush the warm honey over the crispy skin. After a couple of minutes more, test with the meat thermometer to avoid overcooking; it should be 50°F/122°F, and no higher than 52°C/125°F, for perfectly pink duck. Remove from the heat, saving any more juices, cover and rest for 5–10 minutes while you pull, slap and boil the noodles.

To finish the noodles, lightly oil a large work surface area. Take your 8 cocoons out of the fridge, and gently roll each one into a rounded rectangular pancake about 0.5cm (¼in) thick and 15–20cm (6–8in) long. Cover and rest for 10 minutes. Bring a large pan of water to the boil. Place your reserved duck fat and juices in a small mixing bowl.

Once the water starts simmering, pull your first noodle. Using a chopstick or skewer, lightly press into the dough of one rectangle along the middle to make an indentation. This scores the noodle, for ripping after pulling. Then, taking an end in each hand, begin to pull (gently). Once doubled in length, begin to swing the noodle straight up and down (gently), letting the middle of the noodle slap the work surface on the down swing once long enough to do so. Continue to gently slap and pull, until your hands are about 90cm (35in) apart. Place the noodle on the work surface and gently rip apart from either side of the score line, creating a loop. Once you've pulled and ripped two noodles, plop them into the boiling water and cook for 2 minutes, while you pull the next 2 noodles. As they cook, immediately drain and transfer to the duck fat bowl, tossing them in the fat to avoid sticking together. Once all the noodles are cooked and resting in the fat, slice the duck and plate the noodles with the duck on top. Serve the cucumber on the side.

LARGE PLATES 121

Double plum ribs and lime leaf slaw

TUPELO HONEY

One Sunday I came across the Japanese fermented salty plum paste umeboshi at a rather boujie grocery store, brought it home on an ill-thought-through whim, got back to my kitchen and sighed, looking around for inspiration. I gathered together the black plums gently waning on the side and a small jar of probably ancient black garlic paste at the back of the fridge, and the rest is umami, fruity, salty, fragrant history. Cooking the ribs tightly covered in foil first ensures tender moist meat, before removing the foil to caramelise the sticky marinade. Serve with a fragrant zesty slaw to cut through the richness, beers and skinny fries. It is difficult to stop eating these, even when you are full to burst!

Gently cook down the plums in a lidded saucepan with a dash of water, the star anise and four pastes. This should take about 15 minutes. Stir through the oil, honey, sesame whisky, soy, scotch bonnet and vinegar.

Preheat the oven to 150°C/300°F/gas mark 2.

Place the ribs in a roasting tray in which they fit snugly, and cover on both sides with the cooked-down plum sauce. Tightly cover with foil and bake in the oven for 1½ hours.

Meanwhile, make the slaw. Using a mandoline if you have it, shred the cabbages and ribbon the carrot into a mixing bowl, together with the lime leaf, spring onions and coriander. Skin the mango using a swivel peeler, cut the flesh into matchsticks or strips, and add to the bowl.

Combine the honey, lime juice, cider vinegar, sea salt and scotch bonnet, pour into the slaw bowl, and gently toss to coat.

Increase the oven temperature to 180°C/350°F/gas mark 4. Remove the foil from the tray and baste the ribs with any sauce in the bottom of the tray. Return to the oven for 30 minutes or so, until tender and sticky.

Serve the ribs with a large heap of slaw, some salty skinny fries, cold beers and lots of napkins.

Serves 2

For the ribs

4 plums, pits removed

1 whole star anise

¾ tbsp black garlic paste

½ tbsp garlic paste or 2 crushed garlic cloves

1 tbsp ginger paste

½ tsp umeboshi paste

3 tbsp neutral oil

2 tbsp tupelo honey

1 tbsp sesame-infused whisky (see page 241), or 1 tsp sesame oil mixed with 1 tsp whisky

2 tsp dark soy sauce

½ scotch bonnet, finely chopped

Dash of rice vinegar

600g (1lb 5oz) baby back ribs

For the slaw

5cm (2in) wedge of red cabbage

5cm (2in) wedge of white cabbage

1 carrot

1 fresh lime leaf, finely sliced

2 spring onions (scallions), finely sliced lengthways

1 small bunch of coriander (cilantro), leaves torn, stems finely chopped

1 rather unripe mango

1½ tsp tupelo honey

Juice of 1 lime

2 tsp cider vinegar

Large pinch of flaked sea salt

¼–½ scotch bonnet, finely chopped, to taste

Smoked lime and honey chicken

SMOKED HONEY

This smoky, salty, zesty sharing feast is a great one for the BBQ on summer evenings, or for waking up dreary winter days. Serve with a big green salad and cocktails. You can add the smoked element in several ways – through the salt, limes, honey and cooking process. Instructions for home-smoking are on page 238. The black lime powder here really is something else; fusty mummified limes transform into a deeply aromatic magic powder you should use on anything and everything.

Serves 4–6

10–12 skinless, boneless chicken thighs

750g (1lb 10oz) new potatoes, halved

4 corn cobs, sliced into 2.5cm (1in) thick rounds

Glug of neutral oil

800g (1¾lb) thick Greek yogurt or labneh

Tajin seasoning, for sprinkling

1 fresh green jalapeño, finely sliced

2 radishes, finely sliced

Coriander (cilantro) leaves and stems, finely chopped

3 limes, smoked if you like (see page 238)

1 tsp sea salt flakes

Freshly ground black pepper

For the marinade

1½ tbsp black lime powder (see method; from 3–4 black limes)

1 tbsp toasted guajillo chilli powder (see method; from 3–4 dried guajillo chillies)

3 tbsp garlic paste or minced garlic

2 tsp sea salt flakes, smoked if you like

120ml (4fl oz) neutral oil

Juice of 1 lime

2 tbsp white distilled vinegar

3 tbsp smoked honey (see page 65)

To make the black lime powder, blitz the limes in a powerful, sharp-blade blender and pulse until finely powdered. Put in a bowl to one side.

To make the toasted guajillo chilli powder, toast the dried chillies in a dry, hot frying pan until charred, then blitz in the blender and pulse until finely powdered. You can shake out the seeds before toasting for less heat, if you like.

Mix all the marinade ingredients together in a large mixing bowl. Set aside 3–4 tablespoons for the potatoes. Add the chicken thighs to the remaining marinade and thoroughly coat. Cover and refrigerate, preferably overnight, to allow the flavours to infuse the meat.

Remove the chicken from the fridge and grab 2 pairs of metal skewers. Skewer each thigh across 1 pair of skewers, moving onto the second pair when you run out of room; you should get 5 or 6 thighs on each pair.

Cook the potatoes and corn in a pan of salted boiling water until the corn is done, around 5 minutes. Drain both thoroughly and pat dry with kitchen paper. Loosen the reserved marinade with the glug of oil and use to coat the par-boiled potatoes.

Preheat the oven to 200°C/400°F/gas mark 6, or prepare your griddle pan on the stovetop, or your BBQ (grill) on indirect moderate-to-hot heat. Roast, griddle or grill the chicken for 15 minutes, then flip and add the potatoes (on a tray if using your BBQ) and cook for another 15 minutes. Once cooked through and crispy round the edges, remove from the heat. Take the chicken off the skewers and toss with the potatoes and the corn cob rounds.

Apply the yogurt thickly to a serving plate and smooth slightly. Sprinkle with Tajin. Arrange the chicken, potatoes and corn cob rounds across the yogurt. Scatter over the jalapeño and radish slices, chopped coriander stems and leaves. Squeeze half a lime over it all and cut the rest of the limes into wedges to squeeze over the finished plates. Finish with the sea salt, a grind of black pepper and another sprinkle of Tajin and serve with a fresh salad.

Dark beef and aubergine curry

SAFFLOWER HONEY

This wonderfully rich curry is loosely based on 'salli boti', a delicious Parsi dish made with mutton or goat and matchstick potato crisps on top… none of which feature in this. The tenderising yogurt marinade is yet another liberty, after the addition of beef and aubergine, so it is no longer a salli boti per se (par si?), but I urge you to make one of those too another time. This dish can be cooked on the stovetop, but last time I did so my Le Creuset melted a hole in my induction hob, steam clouds stained my ceiling and the bottom of the curry was burnt from lack of stirring. Cooking casseroles, stews and curries in the oven instead allows for an even heat from all sides, requires no stirring, and you can truly forget about it until poking with a fork prompts the beef to fall apart into the unctuous melting aubergine sauce, and it's showtime.

Serves 6–8

800g (1¾lb) braising steak, not pre-diced

100ml (3½fl oz) vegetable oil

1 tsp coriander seeds, roughly crushed in a pestle and mortar

1 bay leaf

1 cinnamon stick

3 large red onions, finely chopped

2 large aubergines (eggplants), or equivalent baby ones, cut into chunky rounds (leave whole if baby)

1 x 400g (14oz) can of chopped tomatoes

2 tbsp safflower or other dark runny honey

2 tbsp white vinegar

1 tsp garam masala

For the marinade

100g (3½oz) plain yogurt

3 tbsp garlic paste

1½ tbsp ginger paste

1½ tsp deggi mirch or Kashmiri chilli powder

1½ tsp fine sea salt

1 tbsp lime juice

½ tsp ground turmeric

1 tsp ground cumin

1 tbsp vegetable oil

Cut the beef into large chunks at least 5cm (2in) across. I don't buy pre-diced because the chunks are often too small and break down too quickly or overcook.

Combine the marinade ingredients in a bowl, add the beef chunks and coat thoroughly. Cover and refrigerate for 6–24 hours.

Heat the oil in a flameproof casserole dish that has a lid, over a moderate/high heat. Sizzle the coriander seeds, bay and cinnamon stick for a couple of minutes until aromatic. Add the red onions and stir frequently for the next 15-plus minutes as they soften and caramelise without going crispy-burnt – add a splash of water if this looks imminent.

Preheat the oven to 150°C/300°F/gas mark 2.

Add the marinated beef and the aubergines to the pan, stir to coat and cook over a moderate heat for about 5 minutes. Add the chopped tomatoes, stir and cook down thoroughly for about 15 minutes. Rinse the tomato can out with a little water and pour this into the dish until the beef and aubergine are covered, then transfer to the oven for about 2 hours, with the lid very slightly ajar to let some steam escape, checking with a fork after 1½ hours to see if the beef is tender yet; be patient. When it is, stir through the honey, vinegar and garam masala. Return to the oven for final 10–15 minutes.

Service with basmati rice, breads, pickles, raita and chutneys. Dip the honeycomb bread from page 160 into the leftovers the following morning.

Goat biryani with jackfruit and lime leaf rice

LINDEN HONEY

I started making biryani regularly after my friend Katie gifted me the Dishoom cookbook, which is an utter joy and a fascinating read, and the pages are now splattered and sticky – always a sign of good fun in the kitchen. This recipe nods to some of my favourite bits from their delicious lamb and jackfruit versions, and introduces some new. Here, the fragrant headiness of fresh lime leaf to infuse the rice, and red cabbage bring a pleasing bite and colour to the goat stew base. I order goat online from Cabrito, but if you can't get hold of it, lamb works perfectly. If you can't get jackfruit, try cauliflower, or halved new potatoes.

Rinse the rice 3–4 times in cold water, submerging then draining, until the water is no longer cloudy and most of the excess starch has been rinsed away, then submerge again and leave to soak. Save the rinsing water for the water/stock element later on in the recipe if you like.

Add the vegetable oil or ghee to a frying pan and caramelise the onion over a moderate–high heat, stirring constantly and adding a dash of water if it starts to catch or stick. This should take around 15 minutes.

Add the whole spices and continue to stir for a minute until fragrant. Introduce the goat to brown it, still stirring thoroughly, for 3–4 minutes. Next add the jackfruit and continue stirring. After 2–3 minutes, add the bay leaves, ginger and garlic pastes, salt and powdered spices. Stir to cook out for about 1 minute.

Add the yogurt, stirring it in thoroughly, and cook the mixture down for 5 minutes. Add the stock or water and turn down to simmer with the lid on and slightly ajar for 50 minutes, stirring occasionally.

Next, stir through the finely sliced red cabbage, the vinegar and tablespoon of honey, and transfer the contents of the pan to a large ovenproof lidded casserole.

Drain the rice and add to a large pan. Add just-boiled water from the kettle to a 7.5cm (3in) depth, stirring through the lime juice and salt. Boil for a few short minutes until robustly al dente, then quickly drain, retaining some moisture. Stir through the finely sliced lime leaf and add to the casserole as a top layer, covering the goat and jackfruit.

Preheat the oven to 180°C/350°F/gas mark 4.

Whisk together the teaspoon of honey and the melted butter or ghee. Pour over the rice, place the lid on the casserole and transfer to the oven for 25 minutes. Remove and stand for 10 minutes before removing the lid, sprinkling over the chopped coriander and serving.

Serves 4–6

3 tbsp vegetable oil or ghee

1 red onion, sliced

1 cinnamon stick

½ tsp black peppercorns

2 cardamom pods

600g (1lb 5oz) diced goat meat

2 x 400g (14oz) cans of jackfruit in brine, drained

2 bay leaves

1 tbsp ginger paste

2 tbsp garlic paste

½ tsp salt

1 tsp deggi mirch or Kashmiri chilli powder

1 tsp ground turmeric

1 tsp garam masala

100g (3½oz) Greek yogurt

150–200ml (5–7fl oz) stock or water

150g (5oz) red cabbage, finely sliced

1 tbsp cider vinegar

1 tbsp linden honey, plus 1 tsp

2 tbsp melted butter or ghee

Fresh coriander (cilantro), finely chopped, to garnish

For the rice

400g (14oz) basmati rice

Juice of 1 lime

1½ tsp fine sea salt

1 lime leaf, finely sliced

Spice cupboard goat with preserved lemons
Ft. Helen's foolproof flatbreads

SAFFLOWER HONEY

Yes, you can do this with lamb! Go forth merrily and without judgement from me. But goat really is delicious, and easy to order online, so go on. Spice-wise, I raided my cupboard like a salivating metal detector, for frankly any flavour that, when imagined clinging to a hot, crispy morsel of meat dripping onto a warm, freshly baked flatbread, made my mouth water. Feel free to ignore the ones I chose below and do the same with your cupboard. You'll regret not doubling the quantities and saving the leftover spice mix in a little pot to cover and cook anything and everything in it – potatoes, sweetcorn, cauliflower, wings, fries.

Helen Graves, queen of flatbreads and live fire cooking, has kindly allowed me to share her no-nonsense recipe for easy, delicious flatbreads, which are the perfect cradle for juicy meat, soaking up all the juices into a wonderful handheld bundle. I've spiced them up with caraway and cumin, because frankly I got a little carried away.

A note on silverskin: goat and lamb shanks, like the back of a rack of ribs, often have a membrane around the outside called the silverskin. It's not a joy to eat, and is a barrier to flavour transfer if you're using a rub or other marinade. If the goat shank has one, use a very sharp knife (I like to use an Opinel pocket knife for this) to carefully remove it by making a cut then pulling back hard, using the knife to release the membrane from the meat as you go.

The goat is seared for a flavourful crust and slow-cooked for melt-in-the-mouth texture. If your meat hasn't given way under a hard stare, it's not ready yet. Tenderness just takes time. You can sear and slow-cook on the BBQ (grill) if you like, even with a little smoke.

Serves 2

1 goat shank
2 tbsp vegetable oil
1 red onion, cut into 8 segments
1 tbsp garlic paste or 3 garlic cloves, minced
½ tbsp ginger paste
1½ tbsp safflower honey
2 preserved lemons, each cut into 6 segments
250ml (8½fl oz) water

For the spice rub
2 tsp cumin seeds
1 tsp coriander seeds
Large pinch of caraway seeds
1 tsp fennel seeds
1 tsp black peppercorns
8 allspice berries
1 tbsp dried oregano
1 tbsp pul biber chilli flakes
1 tsp dried mint
½ tsp dried thyme
2 bay leaves, crushed
¼ tsp garlic powder
½ tsp salt
2 tbsp vegetable oil

Preheat the oven to 200°C/400°F/gas mark 6.

For the spice rub, briefly pummel the cumin, coriander, caraway, fennel seeds, peppercorns and allspice with a pestle and mortar. Mix with the remaining spice rub ingredients, forming a loose paste.

Rub the goat with the spice paste, working it into every nook and cranny. Place the shank in a roasting tray and into the oven to sear for about 30 minutes, turning halfway through. Save any roasting juices from the tray for later. Turn the oven down to 160°C/320°F/gas mark 2.

While the goat is searing, prepare its slow-cooking liquor. Heat the oil in a flameproof casserole large enough to fit the shank, and soften the red onion over a low–medium heat, adding a dash of water if it begins to stick or burn. This should take 10–15 minutes. In a small bowl, stir together the garlic, ginger and honey. After a few minutes, add to the casserole with the preserved lemons, stir and cook for a further 5 minutes or so, then remove from the heat. Once the goat has seared, add it to the dish with the measured water. Pop the lid on and return the goat to the oven for 1½ hours, removing the lid after 1 hour and turning the goat, as the submerged underside will soften more quickly. Add more water if dry, but not too much; you should be left with a thick gravy at the end, not soup. Once the lamb is truly falling apart and the sauce has reduced significantly, remove from the oven and leave to rest in the hot pot with the lid on.

When the lamb first goes in the oven, make the flatbread dough. Toast the caraway and cumin seeds in a hot, dry pan until fragrant, then add to a mixing bowl with the rest of the dough ingredients and combine thoroughly. Knead on a lightly floured or oiled work surface until smooth and springy, then leave to prove in a warm place for an hour or so until roughly doubled in size.

Soak the red onion slices in cold water while you cook the flatbreads.

Knock back the proved dough and divide into 4 balls for large breads or 6 for small. Roll the dough balls flat to about 15cm (6in) then cook for 1–2 minutes in a properly hot, dry pan – seriously, you need to preheat it for 5 minutes at least. Helen and I like to use a cast-iron griddle pan. Cook until a little charred on each side; they will start to puff up when ready.

Brush the flatbreads with some of the reserved roasting juices and stack, covered, in the still-warm oven. Drain the red onion.

To serve, carve a soft spoonful of meat from the shank pot onto a flatbread, squeeze over some lemon juice and add a little parsley or coriander, and slices of red onion.

For the flatbreads (makes 4–6)

1 tsp caraway seeds

1 tsp cumin seeds

250g (9oz) strong white bread flour

3.5g/⅛oz fast-action dried yeast (½ sachet)

1 tsp fine sea salt

1 tbsp olive oil

150ml (5fl oz) warm water

To serve

1 red onion, finely sliced

½ lemon

Small handful of parsley or coriander (cilantro), leaves torn and stems removed

Honey hong shao rou

BUCKWHEAT HONEY

A traditional Chinese recipe with regional variations from Hunan to Shanghai, this is one of those deeply comforting dishes that fills your house with an olfactory hug well before it's on the table. Here I've used honey in place of rock sugar. Embrace the wobbly fat – it's utterly delicious. For vegetarians, I highly recommend using aubergine instead of the pork.

Serves 4

4 spring onions (scallions)

700g (1lb 9oz) pork belly, cut into 2.5cm (1in) chunks, skin on

4 tbsp light soy sauce

2 tbsp dark soy sauce

80ml (3fl oz) Shaoxing rice wine

50ml (2fl oz) buckwheat honey

2 whole star anise

5cm (2in) piece of fresh ginger, finely sliced

2 bay leaves

Slice the spring onions; I like to keep the slices chunky and then halve lengthways. You can slice the green tops finely to garnish the dish later if you like.

Blanch the pork pieces in a large pot, by covering with cold water and bringing to the boil. Once boiling, for the next 2–3 minutes, skim off any gunk that collects on the surface, then drain into a colander and swill out the pot.

Place the pot back over a low heat and add the soy sauces, rice wine and honey, stirring together. Add the spring onions, star anise, ginger and bay leaves. Loosen with about 3 tablespoons of water if necessary to properly coat and get things moving. Keep stirring at a spirited simmer until the spices are fragrant.

Add the pork and enough water to *just* cover the pork, stir thoroughly then bring to the boil. Reduce to a simmer and cook, uncovered, until the pork is tender – about an hour. Scoop up a piece with a spoon and press it against the side of the pot; if the skin and meat break easily, it's ready.

Remove the pork to a warmed plate and reduce the sauce on a high simmer, stirring, until thick and glossy. Reunite the pork with sauce, and serve with steamed rice and greens.

CHAPTER 4

Bakes

140	Garlic fermented honey and miso sourdough	178	Halwa Chebakia
144	Apricot and fennel croissant swirls (a love letter to Pophams bakery)	181	Almond briouats
		184	Loaf cakes
152	Preserved lemon focaccia	188	Custard tart!
154	Saffron and apricot honey buns	191	Honey tarte tatin
156	A tale of two babkas	194	Griddled peach and pistachio pavlova with lavender honey
160	Honeycomb bread – khaliat (al) nahal	197	Honey nut corn cake
163	Brown butter / oak / chestnut madeleines with orange blossom icing	200	Cardamom Basque cheesecake
		203	Rose roast quince with hazelnut meringue and honeyed mascarpone
166	Scauratielli	204	Lemon pollen pie
170	Loukoumades – Greek honey puffs	207	Michelle Polzine's 10 layer honey cake
171	Cut glass paprenjacs	210	Fig leaf panna cotta
174	Baklava		

Garlic fermented honey and miso sourdough

GARLIC FERMENTED HONEY

This crusty umami garlicky loaf has too many delicious elements not to give it a go, even if you're new to the sourdough-at-home game. Dip it in good balsamic and olive oil, serve it with soup, or toast and spread with soft goat's cheese and top with the pickled mushrooms on page 66. To understand a little more about how honey can aid the fermentation in sourdough, see page 33.

Stir together the starter ingredients, pour into an open-top Kilner jar, cover with muslin (cheesecloth) and leave for a day. Discard a third (or use for the focaccia poolish or crumpets on pages 152 and 81), then add the rye flour, water and garlic fermented honey to refresh. Leave overnight; it is ready the next morning.

Preheat the oven to 180°C/350°F/gas mark 4.

In a mixing bowl, mix together the refreshed starter, warm water and bread flour into a scraggy wet dough. Leave the dough to autolyse/rest for 1 hour, covered, in a warm place. If you don't have a warm place, give it a water bath, floating the bowl in a larger mixing bowl. Ideally, you need an internal temperature of 25–28°C/77–82°F.

Meanwhile, cut the tops off the garlic bulbs and place on 2 squares of foil. Drizzle the tops with the olive oil and honey and wrap tightly in the foil. Roast in the oven for 45 minutes or so, until soft. Remove and set aside.

Keeping the dough in its bowl, sprinkle over the salt, wet your hands and fold 16 times, pulling out an edge of the dough from underneath and folding it up into the centre. Turn the bowl as you go, working your way around the dough. Rest, covered, for 30 minutes.

Repeat the folds, then cover and rest for 3–3½ hours.

Repeat the folds. After the 8th fold, carefully add the roasted garlic cloves, squeezing them out of their skins and over the dough. Gently continue with the remaining 8 folds, then turn out onto floured surface to shape; tightening into a round. Lightly mist and generously dust a banneton (if using) with semolina and carefully tip the dough into the basket, top down. Rest for another 1–1½ hours.

Continued

Makes 1 loaf

For the starter

150g (5oz) organic rye flour

300ml (10fl oz) warm water

1 tsp raw honey

To refresh

100g (3½oz) rye flour

100ml (3½fl oz) water

½ tsp garlic fermented honey

To make

150ml (5fl oz) refreshed rye starter (see above)

350ml (12fl oz) warm water

500g (18oz) strong white bread flour (Shipton Mill seeded mix is nice)

12g (½oz) fine sea salt

To bake

2 bulbs of garlic

1 tbsp olive oil

1 tbsp garlic fermented honey (see page 62)

Ground semolina, for sprinkling

To spritz

1 tsp miso paste

1 tsp garlic fermented honey (see page 62)

100ml (3½fl oz) water

Meanwhile, stir together the miso paste, honey and water, and decant into a small spray bottle.

Preheat the oven to as high as it will go. Put a lidded ovenproof casserole, wide enough for the bread and twice the depth, into the oven.

After 15 minutes, allowing the oven and pot to heat up fully, carefully remove the hot casserole from the oven, sprinkle the base with a small handful of ground semolina, to avoid sticking, and gently tip the dough into the pot. I quite like the high stakes of trying to invert the dough into a scaldingly hot pot, but if this is too much drama for you, just place a sheet of baking parchment over the banneton, invert the dough onto it, then lower into the pot holding the parchment edges. If doing so and not risking burning yourself, you won't need the semolina.

Snip or score the top and spritz with the miso/honey water. Put the lid on and bake for 20 minutes, then remove the lid, spritz again and bake for a further 15 minutes. The crust should be crisp and, when turned out, the bread should sound hollow when tapped on the bottom.

Serve sliced with salted butter, cheese and more honey.

Apricot and fennel croissant swirls (a love letter to Pophams bakery)

LAVENDER HONEY

This recipe is a romance story, of which there are several through this book, celebrating brilliant food made by talented people. If you're vaguely in travelling distance of East London, I urge you to visit one of Pophams bakeries for a bacon maple swirl. These huge yet ethereally light croissant swirls have a sweet streaky rasher spiralled through and razor-sharp lamination.

I am far from a professional chef, and just a few black holes away from being remotely well-versed in the art of French pastry. I would usually consider a lie down at the mere thought of attempting it. However, I thought that if I can have a go at lamination and create something delicious at home without losing my mind, then that's something worth sharing for other home cooks.

As a child, whenever we had croissants we would always have them with Bon Maman apricot jam and salted butter. Inspired by this memory, this recipe combines a tart, sticky roast apricot filling, caramelised honey and fennel.

IF YOU ARE NEW TO LAMINATED PASTRY:

Laminated pastry means the pastry dough has been folded around a block of butter, and the two folded together and rolled out again and again to create a many layered sandwich, using careful chilling to avoid melting into one. These many layers give you the swirly layers of croissants, pain au chocolat, and other 'viennoiserie'. In my experience as a beginner to this, the most important elements to remember are: use good ingredients, including quality butter with a high fat content above 82% (French butter is often 82%+); be neat and precise with the folding and cutting; and, most importantly, DON'T be impatient with the chilling time between steps. This is a two-day recipe and sadly can't be rushed. Really, it's a week-long recipe, while you get hold of good flour (probably online), butter and yeast.

However, the swirls do freeze well before the second prove, so you can make a batch, freeze most of them, then transfer one from freezer to fridge the night before, to prove, and bake in time for a mid-morning weekend treat.

Continued

Makes 12

For the dough

275ml (9fl oz) milk

18g (¾oz) fresh yeast

45g (1½oz) caster (superfine) sugar

10g (⅓oz) fine sea salt

250g (9oz) strong white flour, plus extra for dusting

250g (9oz) 00 flour or plain (all-purpose) flour

20g (¾oz) unsalted butter, softened

250g (9oz) cold block of unsalted butter (removed from fridge 30 minutes before using, to make it more malleable)

For the filling

2 punnets of apricots (about 14 large ones)

80ml (5½ tbsp) lavender honey

2 tsp fennel seeds

Large pinch of flaked sea salt

To glaze

2 tbsp warmed lavender honey

DAY ONE

Prepare the dough, butter block and filling. Combine the milk and yeast in a large mixing bowl. Whisk the sugar and salt through the flours to evenly distribute, then stir the dry ingredients into the wet. Add the softened butter and bring together.

Lightly flour a work surface and knead the dough until bouncy and smooth, 10 minutes or so. Check gluten formation with the window pane test; stretch a small scrap of dough and it should allow light through before tearing. Place in the fridge to rest overnight in a lightly floured bowl covered in cling film (plastic wrap).

Next, prepare the butter block. Place the block of butter lengthways on a large piece of baking parchment, at least 45 x 110cm (18 x 43in). The parchment will become the wrapper, helping the butter keep the correct shape. Bash the butter with a rolling pin, working from left to right. Once softened and flattened, fold the baking parchment over the butter to create a 20 x 25cm (8 x 9in) envelope. Continue with the rolling pin until the butter has filled its envelope at a consistent thickness. Chill in the fridge overnight, together with the dough.

Finally, prepare the apricot filling. Heat the oven to 175°C/345°F/gas mark 4. Halve the fruits, remove the stones and arrange in a roasting dish small enough for them to just touch. Drizzle over the honey and scatter with the fennel seeds and sea salt. Cover the apricots with baking parchment and roast for 30 minutes. Remove the parchment, squash the fruit with a fork, then return to the oven for a further 30 minutes until thick, dark and caramelised. Once cooked, mash to a paste with the fork and decant into a container. Store in the fridge overnight.

DAY TWO

Laminate the dough, shape, prove and bake. Remove the dough from the fridge and allow to soften for 10–15 minutes until still firm but rollable. Take the butter out of the fridge and allow to soften while you roll the dough into a 25 x 40cm (9 x 16in) rectangle (twice the length of the butter block).

With a long side of the dough running left to right in front of you, place the butter in the centre of the dough so that the short side is running away from you to make the butter sit flush with the top and bottom of the dough, with 10cm (4in) dough left either side of the butter block. Fold these 10cm (4in) sides in to meet in the middle over the butter and press to join together.

Continued

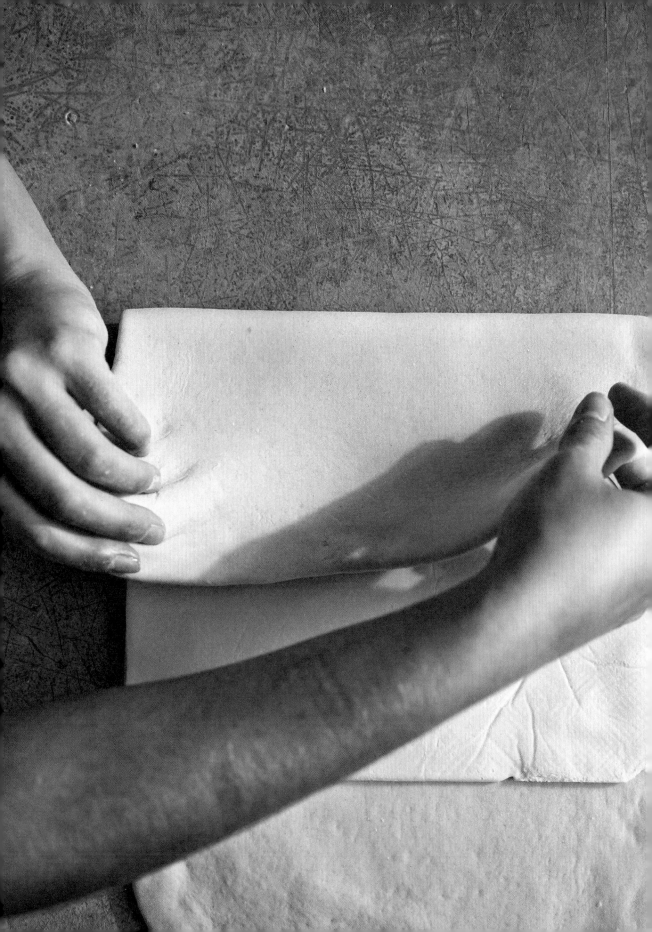

Next, roll the dough up and down along the vertical join, so that the block goes from 25cm (9in) tall to 55cm (21½in) tall, but stays 20cm (8in) wide. Neatly trim the short ends top and bottom with a serrated knife to expose the folded layers, and cut off any overhanging dough.

Turn the dough 90 degrees, then fold the dough in on itself in thirds: left third over centre third, right third over first two thirds, like a formal letter. Wrap your dough letter in cling film and chill in the freezer for 30 minutes.

For the next roll and fold, have the opening of your dough letter to the left. Roll out to 20 x 55cm (8 x 21½in) again, and trim the ends again. Turn 90 degrees, and fold the dough in quarters like a map or leaflet: left quarter into second left, right quarter into second right, then one side over the other. Wrap and chill in the freezer for another 30 minutes if necessary, until firm.

No folding this time, just a roll. With the fold opening facing right, roll out the dough to 30 x 15cm (12 x 6in). Wrap and chill again until firm. The dough is now laminated and ready for shaping.

Roll out the dough until it almost doubles in size – 60 x 25cm (23½ x 10in), with a longer side running left to right. Using a spatula or fork, spread the apricot mixture evenly across the dough. Trim the dough's edges with a serrated knife to create a sharp rectangle, then pick up the long side of dough nearest to you and roll away from you, creating a roulade. Wrap and chill for 30 minutes in the freezer again, to firm up enough for cutting.

Using that sharp, serrated knife, cut the roulade into 12 slices (roughly 4cm/1½in), then arrange on baking trays lined with baking parchment. Cover loosely with lightly oiled cling film, and prove in an airing cupboard or somewhere warm (around 25°C/77°F) until they have almost doubled in size and you can start to see the lamination layers, about 1–1½ hours.

Preheat the oven to 200°C/400°F/gas mark 6.

Remove the cling film and pop the trays in the oven, immediately reducing the temperature to 180°C/350°F/gas mark 4. Bake for 15 minutes until golden brown. As soon as the croissants come out the oven, give them a quick glaze with the warmed honey using a pastry brush. Serve warm.

Preserved lemon focaccia

RAW AND GARLIC FERMENTED HONEY

Preserved lemons are such a treat. Popping open the jar, plopping a tiny fruit onto the chopping board, carefully cutting little jellified slices while their savoury aromatic fragrance wafts through the kitchen. This focaccia sees them tenderly poked into bubbly herbed dough then puddled with garlic fermented honey and good olive oil. Pure sunshine on a baking tray.

Early evening the day before, start by making the poolish. Dissolve the honey and yeast in the warm water, then thoroughly mix through the flour. If your flour seems a little thirsty, add 25–50ml (1–2fl oz) more water; the poolish should be fairly liquid. Cover and leave to prove at room temperature or a little above for 2 hours or so until bubbly and yeasty smelling but before the surface is covered in a mass of bubbles; this is overripe. Now make the dough.

Dissolve the yeast and raw honey in 150ml (5fl oz) of water and stir into the poolish. Sift over the flours and the salt, stirring thoroughly to combine. Knead for at least 5 minutes, up to 10. As this is a wet dough, if you don't have a stand mixer with a dough hook (neither do I), this is easier to do in the bowl with well-oiled hands or a dough scraper, by scooping under the dough and gently pulling it out to stretch it, then rotating the bowl a little and repeating.

Once the dough is smooth and a little bouncy, place in a well-oiled bowl, cover with lightly oiled cling film (plastic wrap) and pop in the fridge overnight.

In the morning, remove from the fridge and allow to come up to room temperature for about an hour. Fold the dough every 20–30 minutes for another 1½ hours, by gently pulling out a corner, folding over on top of the dough, turning the dough and repeating 3 times, then carefully slide the dough so that the folded side is underneath.

Carefully slide the dough out onto a well-oiled baking tray and gently pull out to fill the 4 corners. Cover with lightly oiled cling film and rest for another 30 minutes or so while you assemble your toppings.

Have your lemon segments and flaked salt to hand, and strip the rosemary leaves. Once the dough has rested, discard the cling film and gently poke the lemon segments and pinches of rosemary into the surface. Sprinkle with flaked salt.

Makes 1 loaf

For the poolish

1 tsp raw honey

2.5g (1/10oz) fresh yeast

250ml (8½fl oz) warm water

250g (9oz) strong white bread flour

For the dough

2.5g (1/10oz) fresh yeast

1 tsp raw honey

150–175ml (5–6fl oz) water

About 500g (1lb 2oz) poolish (see above)

150g (5oz) strong white bread flour

100g (3½oz) 00 flour

10g (1/3oz) fine sea salt

Oil, for greasing

For the topping

2 preserved lemons, each sliced into 8 segments, pips removed

Good flaked sea salt

2 sprigs of rosemary

Garlic fermented honey (see page 62), for puddling

A lot of good olive oil

Add a tray of hot water to the bottom of the oven to create steam and preheat the oven to as hot as it will go: 250°C/480°F/gas mark 9, or preferably above. While it's heating up, add warm water and a teaspoon of salt to a spray bottle.

Once the oven is up to temperature, spritz the focaccia lightly with the brine and bake for 15 minutes. Meanwhile, have your garlic fermented honey ready with a teaspoon. After 15 minutes, remove and puddle a little honey into each of the lemon segment dimples. Spritz again with the brine and return to the oven for a further 5–10 minutes until golden brown.

Serve warm, sliced, with more honey and some good olive oil drizzled over, dipped into labneh. Delicious.

Saffron and apricot honey buns

HEATHER HONEY

These buns are delicious. There's no two ways about it. Joyfully yellow, somewhere between a Cornish saffron bun and a Swedish St Lucia bun, but with apricot and citrus peel instead of raisins and sultanas, which never quite get my mouth watering as much as I'd like them to. You can apply a warmed honey glaze straight out the oven, but this makes them anarchically sticky to hold and I love the hard shine of a generous eggwash. Best eaten warm, sliced and toasted with thick salty butter or clotted cream, and a large cup of tea.

Preheat the oven to 150°C/300°F/gas mark 2.

Place the saffron on a small baking tray and toast in the oven for around 10–15 minutes.

Heat the milk in a pan until steaming. Remove from the heat, crumble the toasted saffron into the milk, stir well, and leave to infuse for 10 minutes. Stir in the honey and yeast, whisk briefly and leave for the yeast to activate for a further 10 minutes or so. Finally, stir in the melted butter and the beaten egg. Whisk thoroughly to combine.

Sift the flour and salt into a large mixing bowl. Create a well and pour in the liquid mixture. Stir with a fork or spoon to combine, then tip out onto a lightly floured or lightly oiled work surface (I prefer oiled). Stretch out slightly, tip the chopped fruit on top and knead for 5 minutes until springy and the fruit is evenly worked through. Lightly oil the mixing bowl, place the dough in and cover. Prove for 45 minutes–1 hour at room temperature until it almost doubles in size and partially springs back if poked gently.

Line a baking sheet with baking parchment. Turn the dough out onto the work surface, knock back gently, and portion into 10 evenly sized portions. Roll each into a ball and place on the baking sheet. Cover and prove again for 30 minutes or so until almost doubled in size.

Preheat the oven to 200°C/400°F/gas mark 6.

Brush eggwash over the buns and bake for 20 minutes until deeply golden on top and they have a satisfying hollow tap when knocked on the bottom. Cool on a wire rack.

Makes 10

Large pinch of saffron threads

300ml (10fl oz) milk

75g (2½oz) heather honey

7g (¼oz) fast-action dried yeast (1 sachet)

75g (2½oz) unsalted butter, melted

1 egg, beaten, plus extra for eggwash

500g (1lb 2oz) strong white bread flour

1 tsp fine salt

150g (5oz) mixture of chopped dried apricots and candied citrus peel

Oil, for greasing

A tale of two babkas

WILDFLOWER AND ORANGE BLOSSOM HONEY

I'm sorry, I just couldn't not include a babka. I'm apologising because every cookbook has one these days, and for good reason; they're too fun and too delicious. This braided Jewish cake, also called krantz cake, had been around for several hundred years minding its own business, before a Seinfeld moment in the 90s, a viral Nutella moment in 2013 New York (see Sarah Jampel and Ari Weinzweig), and a London boom thanks to Ottolenghi and Honey & Co's mouth-watering takes. Historically made with leftover challah dough and therefore oil to keep it parve, it is now also often made with butter as a brioche, and with any filling and topping you can think of. A babka loaf is a thing of beauty, a real 'I did that!' moment as you admire it sitting pretty in the tin, then cut through to reveal the intricate swirls of soft dough and rich filling. Why have one when you can have two, so here are two flavour combinations to choose from. You can happily ask yourself, 'Which babka am I in the mood for today?' and get baking.

Pistachio, rose and cardamom

Whisk together the flours, sugar, salt and yeast in a mixing bowl. In a separate bowl, stir together the egg, honey and warm milk. Using a stand mixer with a dough hook or a mixing bowl and wooden spoon, bring together the wet and dry ingredients. The dough should be relatively soft. If a little tough, add some more milk. Knead by hand in the bowl or with the mixer for 5 minutes or so, then begin to add the butter a little at a time, continuing to knead, until the butter is fully incorporated, and the dough is soft, smooth and a little springy. Shape into a ball and place in a bowl under oiled cling film (plastic wrap) to prove overnight in the fridge.

The next morning, remove the dough from the fridge. While the dough is warming up enough to roll out, prepare the filling.

Melt the butter and stir in the honey. In a small mixing bowl, add the butter and honey to the pistachios, cinnamon and cardamom and combine thoroughly. Set aside.

Line a 900g (2lb) loaf (pan) with baking parchment (use a little butter to make it stick), and lightly flour a work surface and rolling pin.

Once the dough is still firm but rollable, roll it out into a 40 x 30cm (15 x 12in) rectangle with a long side nearest to you, and spread with the filling. Using your hands, roll the long side away from you to create a roulade. If the dough has softened considerably by this point, chill in the freezer for 15 minutes or so until firm. Trim the ends to show

Makes 1 babka

150g (5oz) strong white bread flour, plus extra for dusting

150g (5oz) plain (all-purpose) flour

25g (1oz) sugar

Large pinch of salt

7g (¼oz) fast-action dried yeast (1 sachet)

1 large egg, plus 1 egg beaten with a little milk for eggwash

25g (1oz) runny honey

100ml (3½fl oz) warm milk

120g (4oz) unsalted butter, softened

For the filling

50g (2oz) butter

75g (2½oz) wildflower honey

150g (5oz) pistachios, chopped

½ tsp ground cinnamon

½ tsp ground cardamom

a neat spiral, then gently cut the roulade in two, lengthways along the centre. Starting at one end, twist the halves together to form a braid; you can choose to turn out the halves as you go to expose the filling, or keep the plain dough side up. Make sure you braid it gently and loosely, without pulling and stretching, so that the lengths of dough get shorter rather than longer. Tuck the ends under and gently place the braid in the tin. Cover with lightly oiled cling film and leave to prove until almost doubled in size, about 1½ hours.

Preheat the oven to 200°C/400°F/gas mark 6.

Lightly eggwash the top of the loaf and pop into the centre of the oven. Immediately reduce the temperature to 180°C/350°F/gas mark 4 and bake for 25 minutes. Cover the loaf with foil to slow browning, then reduce the temperature to 160°C/320°F/gas mark 2 and bake for a further 25 minutes or until a skewer inserted into the centre of the loaf comes out clean.

While the babka is baking, prepare the syrup. Warm the honey, sugar and water in a pan until foaming but not boiling, remove from the heat and, once cooled slightly, stir through the rose water and lemon juice. Leave to cool completely.

Once the babka is cooked through, remove from the oven and immediately pour over the syrup. Leave in the tin to absorb for 20 minutes or so, then turn out onto a wire rack to cool completely before serving.

For the syrup

50ml (2fl oz) wildflower honey

50g (2oz) white sugar

50ml (2fl oz) water

1½ tsp rose water

Squeeze of lemon juice

Blood orange, almond and dark chocolate

Follow the same dough recipe and method as above. Assemble as above, substituting the filling and syrup.

For the filling, melt the butter and chocolate in a bain marie or a heatproof bowl set over a pan of simmering water, making sure the bowl isn't touching the water. Add the ground almonds, almond butter, runny honey, orange zest and cocoa powder and stir thoroughly. Leave to cool.

For the syrup, warm the honey, sugar and blood orange juice in a pan until foaming but not boiling, remove from the heat and, once cooled slightly, stir through the orange blossom water and lemon juice. Leave to cool completely.

For the filling

50g (2oz) butter

85g (3oz) dark (bittersweet) chocolate

30g (1oz) ground almonds

30g (1oz) almond butter

75g (2½oz) runny honey

Grated zest of 2 blood oranges

30g (1oz) unsweetened cocoa powder

For the syrup

50ml (2fl oz) orange blossom honey

50g (2oz) white sugar

50ml (2fl oz) blood orange juice

1 tsp orange blossom water

Squeeze of lemon juice

Honeycomb bread – khaliat (al) nahal

ACACIA HONEY

These Middle Eastern buns are tender, pillowy joy. Dinky enough to pop into your mouth whole, you'll find yourself reaching for another before you've finished the first. Filled with a little nub of mild cheese and topped with a saffron honey glaze which cracks gently when you take a bite, they really are the best sort of iced bun. Packed cheek by jowl in the tin to create a soft tearing loaf in a honeycomb pattern, the name khaliat (al) nahal, translates from Arabic as 'bee's hive'.

Historically the cheese of choice is labneh, but cream cheese is now often used. Labneh is easy to make, being essentially extra-strained yogurt (mix a generous pinch of salt through Greek yogurt, hang in muslin/cheesecloth over a bowl for 12–48 hours, depending on how wet your yogurt was, until it is dense like playdough and holds the muslin shape well when unwrapped).

If you haven't worked with a wet dough before and you don't have a stand mixer with a dough hook (I don't), be careful not to be tempted to add more flour to make it come together into something easier to handle. The moisture gives you the softness. Rather than making an unqualified, disheartening mess trying to knead on the counter (speaking from experience) I would recommend kneading it in the bowl using a dough scraper. I learned this technique watching one of Jack Sturgess' 'Bake with Jack' YouTube videos. Less messy, simpler, and much more satisfying, ensuring these buns are a joy to make as well as eat, even if you are a less-than-confident baker, like me.

Makes at least 30 buns

240g (8½oz) strong white bread flour

240g (8½oz) plain (all-purpose) flour

3 tbsp caster (superfine) sugar

½ tsp fine sea salt

1½ tsp fast-action dried yeast

2 tbsp vegetable oil

6 tbsp unsalted butter, melted and cooled

2 tbsp plain yogurt

250ml (8½fl oz) milk

1 egg, plus 1 extra egg beaten with a little milk for eggwash

200g (7oz) cream cheese or labneh, chilled

For the glaze

Pinch of saffron threads

30g (1oz) water

60g (2oz) white sugar

2 tbsp acacia honey

Add the flours, sugar, salt and yeast to a mixing bowl and stir to combine. Mix the oil, melted butter, yogurt, milk and egg in a jug (pitcher) until smooth, and pour into the dry ingredients. Stir thoroughly with a wooden spoon until it comes together.

Knead in the bowl for 10 minutes, by scooping down the side with a dough scraper and stretching up, then rotating the bowl slightly and repeating. After 10 minutes of kneading, cover and rest at room temperature for 1–1½ hours until almost doubled in size.

Line 2 loaf tins (pans) or one large springform cake tin with baking parchment. Gently punch back the dough and pinch off into balls smaller than a golf ball. Flatten each ball into a disc a little smaller than your palm. Hold in one hand and use a butter knife to transfer a nub of the cream cheese about 1–2cm (½–¾in) wide into the centre.

Continued

Put the knife down, carefully pinch the outer edges of the dough around the cheese then gently roll into a ball. Arrange the balls snugly in a single layer in the loaf tins or cake tin as you go, in interspaced lines, to create a hexagonal effect. Once you've filled both tins (or the cake tin), cover and set aside to prove for 30 minutes–1 hour, until swollen.

Meanwhile, make the glaze. Toast the saffron in a moderately hot, dry pan until fragrant. Heat the water and sugar in a pan until dissolved, then crumble in the saffron. Remove from the heat and stir in the honey. Set aside to cool completely.

Preheat the oven to 180°C/350°F/gas mark 4.

Brush the tops of the loaves with the eggwash and bake for 20 minutes until golden brown on top. While still hot, pour over the glaze. Leave to soak in and cool for 30 minutes before gently turning the loaves out of their tin/s.

Brown butter / oak / chestnut madeleines with orange blossom icing

OAK HONEY

My relationship with madeleines is having a renaissance. Growing up, madeleines were sticky, gummy, slightly sweaty little cakes to be got out of their plastic wrappers as quickly as possible and shoved into your mouth on holiday at a French campsite. I'm not sure we ever had them fresh from a bakery, or baked them ourselves. Delicious, sure, but a very different delight to a light, toasty madeleine fresh out the oven, with crispy edges. But even those I passed on unawares, as they look a little dry and plain from afar when alongside showier patisserie offerings. Enter two friends, Jess and Alex, and their love for Claire Ptak of Violet Bakery. Jess is brilliant at baking madeleines; her little parcels of freshly baked madeleines came into my life and I've never looked back.

This recipe is based on the standard equal parts butter, egg, flour and sugar that Claire Ptak's recipe uses, with a little nut and brown butter influence inspired by a Saturday morning discovering Flor's brilliant brown butter financiers. If you're finding the elusive hump to be nowhere in sight, check your baking powder is in date and make sure both your tin and batter are well chilled before baking.

Grease your madeleine tin (pan) thoroughly with butter and lightly dust with flour. Freeze for 30 minutes and repeat a few times for peak non-stick ease; once you've finally achieved the acclaimed humps, you don't want to be thwarted getting them out of the tin. Keep the tin in the freezer until it's time to bake.

To brown the butter, melt it in a pan over a medium heat, watching carefully as it starts to bubble and froth. Once it starts to smell slightly nutty and begins to change colour, remove from the heat and swirl. The solids should have browned and gathered on the bottom. Carefully pour off into a heatproof jug (pitcher), leaving the solids behind.

Once the butter has cooled slightly, stir in the honey.

In one bowl, dry-whisk together the (sifted) flours, ginger, salt and baking powder to thoroughly combine. In another bowl, beat together the eggs and sugar until pale and voluminous, using a hand-held or electric whisk. Pour over the butter-honey mixture and continue whisking, then whisk in the dry ingredients.

Continued

Makes 12

100g (3½oz) butter, plus extra for greasing

1½ tbsp oak honey

75g (2½oz) 00 flour, sifted, plus extra for dusting

25g (1oz) chestnut flour, sifted

Large pinch of ground ginger

Large pinch of fine sea salt

¾ tsp baking powder

2 large eggs (100g/3½oz)

100g (3½oz) golden caster (superfine) sugar

For the icing (frosting)

200g (7oz) icing (confectioner's) sugar

2 tbsp orange blossom water

Place a piece of cling film (plastic wrap) over the surface of the batter and chill for at least 1 hour and up to 3 days; overnight is ideal.

Preheat the oven to 200°C/400°F/gas mark 6.

Spoon a scant tablespoon of batter into each madeleine mould. Speaking from experience, it is very easy to over-fill the moulds. I still do it every time, because I am greedy and think bigger will be better, but not so with madeleines; an overweight flabby madeleine is a sad sight.

Pop the madeleine tray into the hot oven and watch them through the window like a hawk. Madeleines cook quickly and these ones will be browner than normal anyway due to, well, all of the non-standard ingredients, so it's important to whip them out before they become too dry; 10–12 minutes should be enough to hump up and spring back to the touch.

While in the oven, make your icing (frosting), if using, by whisking together the icing sugar and orange blossom water. Set aside.

Remove the madeleines from the oven and allow to cool for 2 minutes in the tin, before flipping them in their shells using a fork. Leave until just cool enough to handle, then dip into the icing and eat still warm.

Scauratielli

ORANGE BLOSSOM HONEY

Once known as Magna Graecia to the Romans, the Southern coast of Italy has a rich, ancient Greek heritage dating back to the 8th century BC. At winter solstice the ancient Greeks made honey sweets to welcome the new season. On Christmas Eve, in modern day Cilento on the Amalfi Coast, a stone's throw from the ancient Greek city of Paestum, they still make scauratielli, fried honey pastries in the alpha and omega shapes from the Greek alphabet, signifying the end of one year and the start of another. These ancient sweets are elegant in their simplicity, being equal parts flour and aromatic boiling water – 'scaurare' means 'to boil'. They are cooked like choux paste, shaped then fried and finished with honey. You can flavour the boiling water with whatever you like: limoncello, bay, rosemary and orange blossom are all popular. These days they are often decorated with rainbow sprinkles, but I've gone for a dusting of toasted sweet spices.

Makes at least 20

1 unwaxed orange or bergamot
250ml (8½fl oz) water
Sprig of rosemary
1½ tsp sugar
Pinch of salt
½ tsp Disaronno
¼ tsp orange blossom water
250g (9oz) 00 flour, sifted
Olive oil, for greasing
Vegetable oil, for deep-frying

To serve
1 tsp fennel seeds
1 tsp caraway seeds
¼ tsp ground cinnamon
100ml (3½fl oz) orange blossom honey

Pare off the zest of the citrus into wide strips, using a swivel peeler. Add the water, zest, rosemary, sugar and salt to a large pan and boil for 8 minutes or so.

Meanwhile, toast the fennel and caraway seeds in a dry pan until fragrant. Transfer to a pestle and mortar and gently crush, but don't grind to a powder. Mix with the cinnamon and set aside.

Remove and discard the zest strips and rosemary from the infused water. Stir through the Disaronno and orange blossom water, then add the flour in one go. Using a wooden spoon, vigorously beat together over the heat into a paste. Keep stirring, and remove from the heat once the dough starts to cleanly come away from the sides of the pan.

Turn the dough out onto a work surface lightly oiled with olive oil. Divide the dough into four lightly oiled balls. Take the first ball, and carefully – the dough will be hot – roll out into a long, thin rope, no more than 1cm (½in) thick. Break off a 15cm (6in) length of dough and pinch together 1cm (½in) or so from the ends, to form a rough loop; this is the alpha sign. To make omega, break off a 25cm (10in) length and curve both ends round towards and past the centre, pressing together slightly to join. Repeat to use up the dough rope, then repeat with the remaining dough balls. Arrange the dough shapes on a baking tray lined with baking parchment until ready to fry.

Pour enough vegetable oil into a deep, heavy-based pan to come no more than two-thirds up the sides, and place over a moderate heat. Test-fry one pastry to check the temperature; it should take about 5 minutes to go lightly golden and crisp on the outside. Fry in batches and drain on kitchen paper on a wire rack. Serve warm, drizzled with the orange blossom honey, with the spice mix sprinkled over the top.

Loukoumades – Greek honey puffs

GREEK THYME HONEY

Humble in size but epic in history, these little yeasted doughnuts were supposedly given to victors in the original Olympic Games, and the first recorded recipes date back to a 1226 Arabic cookbook. Loukoumades derives from the Arabic 'luqma' which means mouthful or bite, as they are just the right size to pop into your mouth whole and have been eaten for hundreds of years, from Greece through Turkey (lokma), Egypt, Iraq and Iran (luqma and zalabiyeh) into India. They have a very particular shaping technique that's a little messy and very fun, which is just how I like to cook. Served warm, piled high and drizzled with honey, they are extremely moreish!

Makes about 30

100ml (3½fl oz) warm water

100ml (3½fl oz) warm milk (or water)

1 tbsp sugar

7g (¼oz) fast-action dried yeast (1 sachet)

250g (9oz) plain (all-purpose) flour

¼ tsp fine sea salt

Vegetable oil, for deep-frying

To serve

100ml (3½fl oz) Greek thyme honey

50g (2oz) walnuts, finely chopped

1 tsp ground cinnamon

Mix together the water, milk if using, sugar and yeast, then whisk in the flour and salt to make a smooth batter. Leave covered in a warm place to rise for about 1 hour, until almost doubled and bubbling.

The batter should be fairly wet, but thick enough so that when squeezed in your hand a ball shape pops out from your thumb and forefinger, ready to be scooped off with a spoon.

Pour vegetable oil into a deep, heavy-based pan to come no more than two-thirds of the way up the sides. Place over a moderate heat until hot enough to deep-fry: plop a small amount of batter in; it should bubble immediately.

While the oil is heating, have a wire rack topped with kitchen paper ready to drain the fried loukoumades, a spoon in a little cup of oil ready for shaping, and a strainer ready for transferring the doughnuts onto the paper. If right handed, you will have the bowl of batter to the left of your stove, as close as possible to the pan of oil. Your left hand will be in the batter, and right hand switching between shaping spoon and straining spoon, with the wire rack to the right of the stove.

Once the oil is ready, place your left hand in the batter and scoop the dough into your hand, bringing it up to the side of the bowl, squeezing a golf-ball shape out of your fist, between your thumb and forefinger. Using your right hand, take the spoon out of the oil cup and scoop the ball off your batter hand and gently into the hot oil. Return the spoon to its oil cup before you scoop the next batter ball. Continue until the oil is full of loukoumades, frying away. Fry until golden brown, using your strainer spoon to turn them, then scoop onto the kitchen paper, and repeat with your next batch of batter.

Serve heaped on a plate, drizzled with honey and sprinkled with walnuts and cinnamon. They're amazing with ice cream.

Cut glass paprenjacs

ACACIA HONEY

Paprenjacs are Croatian black pepper, walnut and honey biscuits (cookies). The gingerbread spice mix of cloves, cinnamon and nutmeg pairs well with jammy fruit and cream; I like to sandwich them around a soft scoop of fig ice cream (see page 220), or punch a hole for a ribbon, to hang as festive ornaments. If grinding the walnuts yourself, it's important to stop before all the oils are released and it turns into walnut paste, as this changes the texture of the biscuit. I've made this recipe a little shorter in the shortbread sense, for a flakier, gentle crunch. Paprenjacs are traditionally shaped by pressing into ornate wooden moulds. I have some cut glass punch cups and whisky tumblers that I press into the dough, first dusting their decorative bottoms with a little flour.

Makes about 30

170g (6oz) plain (all-purpose) flour, plus extra for dusting

30g (1oz) rice flour

50g (2oz) sugar

50g (2oz) ground walnuts (not too fine)

¼ tsp ground black pepper

¼ tsp ground cloves

¼ tsp ground cinnamon

¼ tsp ground nutmeg

½ tbsp bee pollen

Pinch of salt

Grated zest of ½ orange

70g (2½oz) very cold lard

70g (2½oz) very cold butter

40g (1½oz) acacia honey

1 egg yolk

Sift the flours into a large mixing bowl and add the remaining dry ingredients and the orange zest. Stir or whisk to mix thoroughly. By hand or with a food processor, add the lard and butter, quickly working the mixture into breadcrumbs. Refrigerate for 10 minutes if the fat starts to melt.

Whisk together the honey and egg yolk, and add to the dry mix. Swiftly combine the dough into a ball and roll out gently to a 0.5–0.75cm (¼–⅓in) thickness. Dust your glassware or moulds with a little flour and press into the dough, cut shapes using biscuit (cookie) cutters or the glass rim. Arrange the biscuits on a baking tray lined with baking parchment and chill for 30 minutes.

Preheat the oven to 125°C/275°F/gas mark 1. Bake the biscuits for 25–35 minutes until cooked through but barely coloured. Transfer to a wire rack to cool.

Baklava

ORANGE BLOSSOM HONEY

Pistachios feature in many of the sweet recipes in this book: babka, pavlova, and now baklava. And for good reason; their smooth, earthy nuttiness is mild and sweet, the perfect foil for pastry, honey and floral waters. I've added pollen and hibiscus to the more traditional mix; hibiscus has a citrus tang and a jewel-pink colour befitting these diamond-shaped pastries, while pollen gives heady floral notes perfect for baklava. Stem ginger shines alongside traditional almond and walnut for a more warming version, perfect for darker days.

Pollen and pistachio with honeyed hibiscus syrup

First, make the hibiscus syrup. Simmer the flowers in the water until vivid pink, strain, then add the sugar until dissolved. Simmer for 10 minutes until thickened, then remove from the heat. When warm but not hot, stir through the honey. Leave to cool completely.

Preheat the oven to 180°C/350°F/gas mark 4.

Mix the dry ingredients for the filling thoroughly before adding the honey, and set aside. Have your melted butter and a pastry brush to hand. Brush a 23 x 30cm (9 x 12in) baking tin (pan) with butter. Place a filo pastry sheet in the bottom of the tin and trim the excess. Use the excess as a guide to trim the same off the remaining stack of filo, so that the sheets will fit the tin. Brush the first sheet with a little of the melted butter, and layer with a second sheet, brushing with more butter. Repeat until you have 8 layers of filo and butter. Add half of the filling and spread in an even layer, gently packed down. Top with 8 more layers of filo and butter, then add the rest of the filling, and the final 8 layers of filo (and butter). Brush the top with butter. Using a sharp, serrated knife (I find a tomato knife works best), cut a diamond pattern through all the layers of filo and filling.

Bake the baklava in the oven until golden brown. Remove from the oven and immediately pour over the cooled syrup, along all the diamond cut lines. Allow the syrup to fully absorb for 8 hours or overnight. Before serving, sprinkle with the extra finely chopped pistachios and bee pollen, then serve.

Makes 1 baking tray batch

200g (7oz) unsalted butter, melted

2 packs of filo (phyllo) pastry (24 sheets)

For the filling

300g (10½oz) unsalted pistachios, finely chopped, plus an extra 2 tbsp for topping

100g (3½oz) bee pollen, plus an extra 1 tbsp for topping

½ tbsp ground cinnamon

Pinch of fine sea salt

2 tbsp orange blossom honey

For the hibiscus syrup

4g dried hibiscus flowers

350ml (12fl oz) water

175g (6¼oz) white sugar

175g (6¼oz) orange blossom honey

Almond, walnut and stem ginger with orange blossom honey syrup

Dissolve the sugar in the water, add the ginger syrup and simmer for 10 minutes until thickened. Remove from the heat and, when warm but not hot, stir through the honey and orange blossom water. Leave to cool completely.

Mix the filling ingredients and assemble and bake as for the pollen and pistachio recipe opposite, sprinkling with the extra finely chopped almonds in place of the pistachios and pollen.

Makes 1 baking tray batch

200g (7oz) unsalted butter, melted

2 packs of filo (phyllo) pastry (24 sheets)

For the filling

250g (9oz) almonds, finely chopped, plus 2 tbsp for topping

100g (3½oz) walnuts, finely chopped

4 balls of stem ginger, finely chopped

1 tsp ground cinnamon

Large pinch of fine sea salt

For the orange blossom honey syrup

125g (4½oz) sugar

350ml (12fl oz) water

50g (2oz) stem ginger syrup (from the jar)

175g (6¼oz) orange blossom honey

2 tbsp orange blossom water

Halwa chebakia

ORANGE BLOSSOM HONEY

These beautiful Moroccan treats are often eaten after a day's fasting at iftar during Ramadan, or on other special occasions. I came across them thanks to the work of food writer Christine Benlafquih, who has spent over 20 years documenting Moroccan cooking and generously sharing her recipes online for everyone to try. Christine has kindly allowed me to include two of her Moroccan pastry recipes in this book – Halwa chebakia and Almond briouats (see page 181). Chebakia are intricately shaped into highly ornamental flowers or braids, but the technique is quick and satisfying once you find your rhythm. You can use either a fluted pastry wheel or a special chebakia biscuit cutter to achieve the characterful crinkled edges. Not only are they exquisite to look at, they're orange blossom-flavoured, deep-fried, drenched in honey and sprinkled with sesame seeds – what's not to love?

Crush the mastic gum grain (if using), sugar and saffron in a pestle and mortar. Add the sesame seeds and grind until the sesame oil is released and creates a moist paste.

Mix thoroughly with the flour and remaining dry ingredients in a mixing bowl. Add the egg and all the remaining ingredients and bring together to form a dough. It should be stiff but workable. Knead for 5–10 minutes, until it is smooth and has some spring, divide into two portions, then leave to rest under cling film (plastic wrap) for 15 minutes.

Take one dough portion and roll it out on a lightly floured work surface into a square about 2mm (1/12in) thick. If you don't have a chebakia cutter, use a fluted pastry wheel to cut out twelve 10cm (4in) squares, then use the pastry wheel to make at least 4 cuts, evenly spaced across the square, like stripes, stopping 1cm (½in) short at either end of the cut. Loosely cover the cut squares in cling film while you are folding each biscuit, to stop them drying out.

Continued

Makes 24

1 mastic gum grain (optional)

Large pinch of caster (superfine) sugar

¼ tsp saffron threads, crumbled

75g (2½oz) sesame seeds, preferably unhulled, toasted

250g (9oz) plain (all-purpose) flour, plus extra for dusting

¼ tsp baking powder

¼ tsp salt

½ tsp ground cinnamon

¾ tsp ground anise

¼ tsp ground turmeric

1 medium egg

2 tbsp melted unsalted butter

2 tbsp olive oil

2 tbsp apple cider vinegar

2 tbsp orange blossom water

½ tsp fast-action dried yeast, dissolved in 2 tbsp warm water

To fry and finish

Vegetable oil, for deep-frying

250ml (9fl oz) orange blossom honey

1 tbsp orange blossom water

Sesame seeds, for sprinkling

To flower fold: pick up a cut square using one hand. Using your other hand, thread your middle finger like a needle through the cuts in the biscuit, alternating under and over. Using your non-threaded hand, pinch together two corners and hold the pinched end. This becomes the centre of your flower. Raise your threaded hand above the pinched end, so that the dough drops along your finger and off over the pinched end. Turn out each strip as it falls off your finger, either side of the pinched end; these are the petals arranged around the centre. You should be left with a rosette of folded strips between the remaining two square corners, which you can pinch to tapered points if you like. Arrange on a baking sheet and cover with cling film until ready to fry.

Repeat with the second portion of dough.

Pour enough vegetable oil into a deep, heavy-based pan so that it comes no more than two-thirds up the sides, and pour the honey into a small saucepan next to it. Heat both to a medium-high heat (the oil should be around 175°C/350°F, if you have a thermometer). The honey will probably heat more quickly – once it begins to foam slightly, turn the heat off and stir through the orange blossom water. Make sure the honey remains hot while working through the pastries in batches, reheating from time to time if necessary.

Don't be tempted to heat the oil too high either; the pastries should fry gently for 6–10 minutes, and become crisp all the way through without going over-brown.

Fry the chebakia in batches, removing with a slotted spoon or 'spider' to drain excess oil before immediately placing them into the hot honey.

Allow the fried chebakia to soak in the honey for 5 minutes or so, until they've turned a rich and glossy amber colour. Remove, sprinkle with sesame seeds and allow to cool properly before serving. They will store for up to a month at room temperature in an airtight container.

Almond briouats

EUCALYPTUS OR ORANGE BLOSSOM HONEY

This is Christine Benlafquih's recipe for the gorgeous little crunchy almond pastries that are traditional, popular and iconic in Morocco. They are made with warqa pastry – which is similar to filo (phyllo) but thinner and less fragile – folded around a ball of almond paste to make a tiny bite-sized triangle. You can substitute with filo by doubling up the sheets and using plenty of melted butter to avoid breaking while folding. Christine recommends using good-quality orange blossom water and a mild honey like eucalyptus. I like to use orange blossom honey to double down. I've specified blanched almonds here, to speed things up.

Makes 25 small pastries

250g (9oz) warqa or 500g (1lb 2oz) filo (phyllo) pastry

30g (1oz) unsalted butter, melted

1 egg yolk, lightly beaten

Vegetable oil, for frying

600g (1lb 5oz) mild-flavoured honey, such as eucalyptus or orange blossom

½–1 tbsp orange blossom water

Sesame seeds, to finish (optional)

For the almond paste filling

200ml (7fl oz) vegetable oil

500g (1lb 2oz) blanched almonds

1 mastic gum grain

175g (6¼oz) granulated sugar

¼ tsp ground cinnamon, or more to taste

Pinch of salt

60g (2oz) unsalted butter, softened

40ml (1½fl oz) orange blossom water

First, make the almond paste. Heat the oil in a pan over a moderate heat and fry half the almonds in batches until golden. This should take 5 minutes per batch so don't have the oil too hot. Drain on kitchen paper and leave to cool.

Grind the mastic gum grain with a pinch of the sugar, using a pestle and mortar.

Blitz the remaining (un-fried) almonds in a food processor together with half the remaining sugar and the mastic gum mix. Repeat with the fried almonds and remaining sugar, then combine the two almond mixtures.

Add the cinnamon, salt, butter, and orange blossom water, and stir into a soft, pliable paste, kneading to ensure it is thoroughly combined. Add a touch more cinnamon or orange blossom water as necessary, to taste. Shape the paste into cherry-sized balls.

Cut the warqa or double filo sheets into long strips, 5cm (2in) wide, using a pizza cutter or pastry wheel. Brush the centre of a strip with a little of the melted butter and place a ball of almond paste at one end. Take a corner of the pastry at the filling end and pull it diagonally over the filling to the other side, enclosing it in a triangle. Continue to fold the pastry around the filling by flipping the bottom corner up to the left, then again to the right, and so on. Press the filling gently to fill out the triangle during the first few folds. Fold the triangle 4 or 5 times; this should take you to the end of the strip. Trim off any excess pastry rather than folding, to avoid too many layers of pastry. Brush the end flap with egg yolk to help seal and tuck into the fold.

Continued

Pour enough vegetable oil for deep-frying into a large, heavy-based pan, so that the oil comes no more than two-thirds of the way up the sides, and place over a medium heat. In an adjacent pan, heat the honey and stir through the orange blossom water. Turn off the heat when it starts to foam. Arrange a wire rack over a baking tray to one side.

Once the oil is ready, fry the briouats in batches, turning and moving them until light golden brown, 5–7 minutes; any quicker and your oil is too hot. Use a slotted spoon to transfer the fried pastries directly from the oil to the hot honey. The pastries should soak for the same time that they fry, so you can have a batch frying and a batch soaking in honey simultaneously. Transfer the soaked pastries to the wire rack and scatter with sesame seeds, if you like, before they cool.

You may need to reheat the honey during this process or add more honey; you need enough to coat the pastries and it must be hot.

Serve once cool enough to touch, and leave for an hour or longer to cool thoroughly before storing in an airtight container at room temperature for up to a month. They freeze well.

Loaf cakes

THYME AND ORANGE BLOSSOM HONEY

If you need a little more joy in your life, and by joy I mean homemade cake, loaf cakes are for you. Every time I bake a loaf cake I rejoice in how easy and no-fuss they are to put together, yet so satisfyingly tasty, sliceable and storable. A cake for real life. It's probably a good thing that I forget all this for weeks at a time before making another one, because otherwise I might roll down the street. What's more, loaf cakes are just the right size to be entirely acceptable for baking if you live on your own, like I do; big round cakes tend to stay sulking in your cookbook until there's a birthday and you have an excuse to make one for friends. You need no excuse to make these.

Almond, olive and raspberry

Line a loaf tin (pan) measuring about 24 x 12cm (9 x 5in) with baking parchment or, preferably, a light buttering and (caster/superfine) sugaring. Strip the leaves from the lemon thyme and add to a mixing bowl with the sugar and citrus zest. Mix with your hands, working the leaves and zest into the sugar to release the essential oils. It will take on colour and become aromatic. Lightly dust the raspberries in 1 tablespoon of the flour and set aside.

Preheat the oven to 180°C/350°F/gas mark 4.

In a bowl, whisk the eggs into the yogurt, then whisk in the oil and honey. Sift over the remaining flour, the baking powder and salt, and add the ground almonds and fragrant sugar. Mix quickly then pour two-thirds into the loaf tin. Arrange the raspberries over the top, then add the remaining batter and sprinkle over the flaked almonds and extra tablespoon of sugar.

Bake in the oven for 25–35 minutes until a thin metal skewer inserted into the centre of the cake comes out clean.

Makes 1 loaf cake

3 sprigs of lemon thyme (1g leaves)

120g (4oz) caster (superfine) sugar, plus an extra 1 tbsp for topping

Grated zest of ½ orange and ½ lemon

100g (3½oz) raspberries

125g (4½oz) plain (all-purpose) flour

2 eggs

75g (2½oz) plain yogurt

75g (2½oz) extra virgin olive oil

2 tbsp thyme honey

1 tsp baking powder

Pinch of sea salt

75g (2½oz) ground almonds

Small handful of flaked (slivered) almonds, toasted

Orange, yogurt and cardamom

Preheat the oven to 180°C/350°F/gas mark 4. Line a 24 x 12cm (9 x 5in) loaf tin (pan) with baking parchment.

Beat together the butter and sugar until pale and creamy, then add the eggs one at a time, mixing well between. Add the honey and yogurt and mix slowly until combined.

Sift the flour with the bicarbonate of soda, then gradually add to the butter and sugar mix. Stir through the orange zest and cardamom seeds.

Spoon the mixture into the prepared tin and gently level the top with a spoon. Bake on a middle shelf of the oven for 50–55 minutes, until a thin metal skewer inserted into the centre of the cake comes out clean. Take out of the oven and leave in the tin while you prepare the drizzle.

Mix the orange juice with the sugar and honey, then prick the cake all over, using a skewer, and gently spoon the drizzle over the cake. Leave the cake to cool in its tin; it will absorb the honey and orange juice.

Remove from the tin and slice. Store any leftovers in an airtight container for 3–4 days.

Makes 1 loaf cake

225g (8oz) unsalted butter, at room temperature

100g (3½oz) caster (superfine) sugar

3 eggs

75g (2½oz) orange blossom honey

70g (2½oz) plain yogurt

225g (8oz) self-raising flour

¼ tsp bicarbonate of soda (baking soda)

Grated zest of 1 large orange

Seeds of 6 cardamom pods, lightly crushed

For the drizzle

Juice of 1 large orange, strained

25g (1oz) caster (superfine) sugar

25g (1oz) orange blossom honey

Custard tart!

CLOVER OR HEATHER HONEY

I am a custard tart generalist. I have fond affection for the anaemic little foiled tarts with very slapdash applications of nutmeg and soggy pastry that are readily available in most big supermarkets, just as much as I love a perfect pastel de nata. London restaurant Quality Chop House are rightly famous for their delicious honey custard tart, which has inspired this one. Using whipping cream as well as, or instead of, double (heavy) cream makes for the daintiest, most tender custard wobble that just about holds itself together when sliced, then immediately melts in the mouth. I like to slightly caramelise the top, which brings out the smoked honey note wonderfully, and I've added familiar custard-loving spices to the pastry case, together with orange zest. All in all it is a Very Impressive Dessert and absolutely worth the wobbly car ride to take it to a dinner party.

If the thought of making a pastry case from scratch means you rarely end up making tarts, buy a block of pre-made shortcrust and add the sugar, spices and orange zest to it (see the hack on page 190), though I promise you it's very quick and easy to make. In fact it comes out better the less time you spend throwing it together; it doesn't like to be overworked. However, if shop-bought is easier for you, go forth. A slightly cheated custard tart is infinitely better than no custard tart. Make like QCH and serve with a stem ginger crème fraîche.

Makes a 20cm (8in) tart

For the custard filling

300ml (10fl oz) whipping cream

200ml (7fl oz) double (heavy) cream

1 strip of pared lemon zest (using a swivel peeler)

1 bay leaf

6 egg yolks (save the whites for the meringue on page 194)

150g (5oz) honey: 1 tbsp lightly smoked honey (see page 65), the rest your honey of choice; I like clover, or heather, like QCH

For the pastry (makes enough for 2 cases; freeze half for later)

Grated zest of 1 orange

75g (2½oz) caster (superfine) sugar

250g (9oz) plain (all-purpose) flour

2 tsp ground ginger

2 tsp freshly grated nutmeg

125g (4½oz) cold butter

½ tsp vanilla extract

1 egg

For the pastry, rub the orange zest into the sugar. Mix the flour, sugar, ginger and nutmeg in a large bowl. Grate the cold butter into the dry ingredients and rub together to a breadcrumb consistency. If the butter becomes soft or starts to melt, pop the bowl in the fridge for 10 minutes before continuing. Add the vanilla and egg, then quickly bring it together into a ball. If it needs a little more moisture, add a drop of water. Wrap the ball in cling film (plastic wrap), squash the top down to flatten slightly and chill for 30 minutes.

Cut the dough in half and freeze one half for another time. Roll the chilled pastry out to 3mm (⅛in) thick and large enough to line a 20cm (8in) tart tin (pan) with excess, then carefully drape into the tin and trim off any excess around the edge beyond 3cm (1¼in). Chill in the fridge for at least 30 minutes, even overnight.

Preheat the oven to 200°C/400°F/gas mark 6.

Continued

Remove the tart case from the fridge. Cut a square of baking parchment large enough to line the pastry with excess. Scrunch into a ball then pull out flat; the softened crinkles will make it much easier to gently line your pastry. Place the crumpled sheet in the tart case, and fill with baking beans (pie weights), carefully rolling them to the edges. Bake for 15–20 minutes, then remove the beans and parchment and bake for another 15–20 minutes until very lightly golden.

Do not let the pastry significantly colour, as this burns off the flavours. Set aside, and turn the oven down to 160°C/320°F/gas mark 2.

Warm the creams, lemon zest and bay leaf to 50°C (122°F) on the stove (I use my BBQ meat thermometer for this... after careful washing) and remove from the heat. I often get distracted while doing this and overcook the cream; don't.

In a bowl, whisk the egg yolks and honey(s) together, then whisk in the warm cream until it starts to thicken slightly. It will foam considerably – hand-whisk to reduce the amount of foam, as this part will be discarded. Pass through a fine sieve into a jug (pitcher) and leave for a few minutes. Skim off the foam. At this point it can be stored in the fridge until ready to use, or used immediately (it should be cooked from room temperature).

Place the tart case on a large, rimmed baking sheet. Pour the custard filling into the case and bake in the oven for 45 minutes–1 hour; it should still be liquid in the centre (extreme wobble). Preheat the grill (broiler) to high, and grill (broil) the tart for 5 minutes until a nice leopard-spot brûlée forms around the edge but the tart still vigorously wobbles. Remove, leave to cool on a wire rack for 1 hour, then serve.

SHOP-BOUGHT PASTRY HACK

Roll out the pastry block to 5mm (¼in) thick. Evenly scatter over the sugar and orange zest, and use a sieve to distribute the spices. Fold the pastry in half like a book, with the extra ingredients in the middle. Roll the pastry out and line the tart tin as above, then chill in the fridge for 30 minutes.

Honey tarte tatin

APPLE BLOSSOM HONEY

When is an apple not an apple? When it elopes with pastry, sugar and butter, shedding any pretence of health to transform into a seductive toffee caramel cloud, and goes by the name of tarte tatin.

I began my gardening journey volunteering in the Kitchen Garden at Raymond Blanc's Le Manoir Aux Quat' Saisons, where he and his team of gardeners planted a magnificent orchard of 150 apple varieties, at least six of which are perfect for Maman Blanc's tarte tatin. I also have fond memories of making tarte tatin with my mum, from the back of a Sainsbury's recipe card. Très chic.

Sadly few of Raymond's varieties are likely to be found in your supermarket, but I recommend Cox's Orange Pippin, Braeburn, Granny Smith, or a combination of all three. I like my apples halved, with skin on, for a little more bite and flavour. Using entirely honey instead of sugar is very luxurious, bringing a wonderful floral aroma. Honey caramelises more quickly than sugar, so the apples still retain some of their tart freshness under the toffee glaze. I'm with Felicity Cloake on the choice of pastry – either shortcrust or puff pastry work just fine. We always used shortcrust growing up and Raymond uses puff; both are delicious.

Makes a 20cm (8in) tart

3 Cox's apples

3 Granny Smith apples

Juice of ½ lemon

1 sheet of puff or shortcrust pastry (or make your own)

150g (5oz) apple blossom honey

50g (2oz) butter, cubed

Preheat the oven to 150°C/300°F/gas mark 2.

Peel the apples if you like, but I like to keep the skins on. Core them and cut into halves. Brush the cut sides with the lemon juice – this is traditionally to stop browning, but they will go brown in the oven. I still do this because I like the subtle zing the lemon brings to the apples. Arrange the apple halves on a baking tray, rounded side down, and pop in the oven for 15–20 minutes while you prepare the rest of the tarte.

Using a 20cm (8in) ovenproof frying pan as an outline guide, cut a circle in your pastry sheet. Return the pastry to the fridge.

Heat the honey in the ovenproof pan on the stove until fiercely bubbling. Keep an eye on it while it bubbles and foams up. It will start to darken, and once it does so, remove from the heat and add the butter. Once it has melted into the honey, whisk together. Leave to cool on the side, whisking intermittently to keep the honey and butter combined.

Continued

Once the honey butter caramel has cooled sufficiently so as not to slide around the pan, and is quite solid but still pliable, take the pastry round out of the fridge and hold it up by the edge. Work your way round the edge, pinching slightly; this will pull out the edge into a thin frilly rim, to be tucked around the edges of the tarte later. Remove the apples from the oven and raise the oven temperature to 200°C/400°F/gas mark 6.

Carefully arrange the apple halves in the caramel pan, rounded side down, tightly packing and overlapping them. Place the pastry over the top. Using the end of a fork or a spoon handle, work your way around the edge, lifting the apple and caramel (it should be hard enough to pull upwards from the pan) and tucking the excess pastry frill down underneath. Prick the pastry all over with a fork, slice a hole in the middle for steam to escape, and bake in the oven for 20 minutes, until the caramel is dark and bubbling around the edges, and the top of the pastry is deep gold in colour.

Leave to cool on the side for 5–10 minutes, for the caramel to solidify a little. If you skip this, the caramel will be too runny when you flip the tart. If you leave it too long, the tart may be fully stuck to the pan (which you can reverse by heating gently on the hob/stovetop before you attempt flipping).

When you're ready, place your serving plate on top of the pan and carefully invert both. Serve with cream or crème fraîche.

Griddled peach and pistachio pavlova with lavender honey

LAVENDER HONEY

Meringue, the kind that's light and crisp on the outside and softly marshmallow on the inside (not the soft one you blowtorch), is one of my favourite desserts to make. There is a photo of a very young me on top of my parents' dining table at Christmas, covered head-to-toe in meringue, berries and cream, having snuck in to hoover the leftovers once the adults had dozed off on the sofa. There isn't a time of year that I'm not inclined to make this greatest of messy-glamorous desserts. Dramatic yet simple, beautifully sculptural yet relaxed and decadently messy to eat, there is a pavlova for every season. It is never not a good time for pavlova.

The flavour for this pavlova stems from a blissful holiday breakfast, watching swallows and swifts soar in the morning air – peaches and honey on Greek yogurt, sprinkled with pistachios, their bright pink skins looking like rose petals. You can stick with rose and use petals and rose water if you like, but I've gone for a little lavender here. I love to add nuts to meringues; it balances the sweet with a roasted savoury nuttiness, making them even more irresistible and complementing any fresh, juicy fruit or smooth, rich chocolate, and of course dollops of whipped cream.

Serves 6

For the meringue

6 large egg whites (save the yolks for the custard tart on page 188)

Large handful of hulled roasted pistachios (unsalted)

360g (12½oz) golden caster sugar

Scant 1 tbsp lavender honey

To serve

3 ripe peaches

1 tbsp sunflower oil

200g (7oz) whipping or double (heavy) cream

200g (7oz) Greek yogurt

3 tbsp lavender honey

1 tbsp dried lavender heads, crumbled

Preheat the oven to 220°C/425°F/gas mark 7.

Plop the egg whites into a large mixing bowl and have an electric whisk ready.

Roughly chop half of the pistachios and finely chop the other.

Spread the sugar over a baking tray lined with parchment, and bake for 8 minutes, until just beginning to melt around the edges.

Meanwhile, whisk the egg whites until bubbly but not foamy, and have the honey to hand.

Remove the sugar from the oven and turn the temperature down to 160°C/320°F/gas mark 2. Drizzle the honey over the sugar and allow to warm up for a minute.

Slowly shake the sugar/honey into the egg whites, pausing to whisk thoroughly after roughly every 2 tablespoons or so. Once the egg whites are glossy and have formed stiff peaks, stop whisking and scatter over half of the roughly chopped and finely chopped pistachios. No need to mix, the pistachio will gently swirl in as you transfer to the baking trays and shape.

Continued

If making one large sharing pavlova, use a spatula to turn all of the meringue mix out onto a baking tray lined with baking parchment, and spread to the desired size with as few movements as possible. If making individual pavlovas, use a large serving spoon to carefully carve big dollops onto lined tray(s).

Pop the baking tray(s) into the oven and immediately turn the temperature down to 120°C/250°F/gas mark ½. Cook until hard on the outside, about 1–2 hours depending on the size of your meringue. Turn the oven off and open the oven door, then leave to cool inside the oven.

While the meringue is cooling, halve the peaches, discarding the stones (pits), and heat a griddle pan to a medium heat. Lightly brush the pan with the oil to stop them sticking. Griddle for a few moments on each side until nicely charred, then remove to a plate to cool slightly.

Remove the meringue from the oven. Whisk the cream to very soft peaks, then whisk in the yogurt until spreadable. Swirl through 1½ tablespoons of the honey. Keep the cream chilled until the meringue has cooled.

Top the meringue with the yogurt cream and peaches. Drizzle with the remaining honey, and scatter over the rest of the pistachios and the lavender heads, and serve.

Honey nut corn cake

TUPELO HONEY

Somewhere between a cake and a cornbread, served warm straight from the skillet, this simple cake is indeed inspired by one of my favourite childhood breakfast cereals, and is just as moreish. Serve with a dollop of crème fraîche and a cup of tea.

Preheat the oven to 180°C/350°F/gas mark 4. Place a seasoned 23cm (9in) cast-iron skillet/frying pan in the oven to warm up.

Finely chop half of the peanuts. Break up the remaining peanuts into halves and set aside.

Cream the butter with the sugar, add the eggs one at a time, then mix in the peanut butter, honey and buttermilk.

Mix the flour, cornmeal and baking powder together in a bowl, make a well in the centre and pour in the wet mixture. Stir together with a fork, then fold in the finely chopped peanuts.

Carefully remove the hot skillet from the oven and pour in the cake batter. Sprinkle over the peanut halves and return to the oven until deep gold in colour, about 25 minutes.

Makes a 23cm (9in) cake

100g (3½oz) roasted peanuts
150g (5oz) butter, softened
100g (3½oz) golden caster sugar
3 large eggs
1 tbsp peanut butter
50g (2oz) tupelo honey
100g (3½oz) buttermilk
75g (2½oz) plain (all-purpose) flour
75g (2½oz) cornmeal
1½ tsp baking powder

Cardamom Basque cheesecake

ORANGE BLOSSOM HONEY

If you have somehow managed to miss the Basque cheesecake glow-up of recent years, I hereby extend its reign with this new take, so you haven't missed the party. One to elicit a surprised 'oooh' from otherwise cheesecake sceptics like me, this is a world apart from an anaemic, shiny slice of cold cloy. I tested this flavoured version on a dank grey August afternoon in the basement staff kitchen of Kew's Palmhouse. It came out perfectly unscathed and resplendently souffléd from a delightfully retro 1970s Starlight oven; a testament to how easy it is to make. Switch up the cream cheese; try mascarpone, labneh, requesón. If you can, infuse the cream with the cardamom the night before. Pairs well with a black coffee the morning after.

Makes a 20–23cm (8–9in) cheesecake

3 or more plump brown cardamom pods

480ml (1 pint) double (heavy) cream

Butter, for greasing

900g (2lb) cream cheese

180g (6oz) caster (superfine) sugar

3½ tbsp orange blossom honey

6 large eggs

1 tsp vanilla extract

45g (1½oz) plain (all-purpose) flour

½ tsp fine salt

Peel the husks off the cardamom pods and tease apart the seeds. Toast both in a dry pan until fragrant, then tip into a pestle and mortar and bash and grind until it won't break a molar. If you do have the foresight and patience to infuse overnight, stir the husks and ground seeds into the cream and keep in the fridge until you're ready to bake. If you don't infuse the cream overnight, instead *very* gently heat the cream in a pan and stir though the cardamom. Keep on a *very* low heat, stirring occasionally, while you do everything else. Allow to cool and strain before using.

Preheat the oven to 200°C/400°F/gas mark 6. Butter a springform cake tin (pan), at least 20cm (8in), and line with baking parchment, leaving plenty (and I mean plenty) of raggedy overhang.

Using a stand mixer or hand-held electric whisk, beat the cream cheese, sugar and honey together on a medium speed until smooth, taking care to scrape in any rim flingings with a spatula as you go.

Add the eggs one at a time, thoroughly whisking before adding the next. Slowly pour the strained cream into the bowl while still whisking. Dribble in the vanilla and whisk through to finish. Your bowl will be alarmingly full and sloppy. Stop whisking briefly. Sift over the flour and sprinkle on the salt, then scrape down any final rim flingings and whisk in slowly, until smooth.

Carefully pour your laden bowl of batter into the parchment-lined tin. It will require slow and tender manoeuvring into the oven. Watch it puff up gloriously into a huge soufflé, and remove once the top is a nice chestnut brown but the centre still has a lively wiggle, which may take 45 minutes–1 hour. Transfer to a wire rack and wait for a dramatic deflation, then remove the spring rim and let cool completely. Serve at room temperature, *not* chilled.

Rose roast quince with hazelnut meringue and honeyed mascarpone

APPLE BLOSSOM HONEY

Roast quince is one of those syrupy, sticky delights that can be cold spooned from the fridge or reheated for a week or so for all sorts of treats: on porridge, with cheesecake, next to ice cream or custard. Here I've paired it with a simple honeyed mascarpone, and I'd highly recommend using the pavlova recipe on page 194 to make hazelnut meringues (simply swapping pistachio for hazelnut) to give a delicate toasted nut crunch to your spoonful of warm, soft quince and cold, smooth mascarpone. Rose, apple blossom honey and cardamom bring out the rich natural perfume of the quince: floral, sweet and spicy. The quantities here make extra syrup, to pour off into a bottle and save for cocktails. Rose roast quince fizz? Yes please.

Serves 6

1 quantity of meringue (see page 194, substituting pistachios for hazelnuts)

50g (2oz) apple blossom honey

500g (1lb 2oz) mascarpone

For the quinces

4 plump cardamom pods

Pared zest of 1 lemon (in strips, using a swivel peeler)

2 bay leaves

1 tbsp rose petals

300ml (10fl oz) water

100g (3½oz) apple blossom honey

80g (3oz) caster (superfine) sugar

2 quinces

Prepare the meringue following the instructions on pages 194–196, using hazelnuts instead of pistachios, and making individual pavlovas or meringues. This (and/or the quince) can be done in advance.

For the quince, bash the cardamom pods briefly to split them open slightly, then toast in a hot, dry pan until fragrant. Add to a small saucepan with the lemon zest, bay, rose petals and water. Gently simmer with a lid on for 10 minutes. Strain, return the liquid to the pan and add the honey and sugar. Stir to dissolve then set aside.

Preheat the oven to 160°C/320°F/gas mark 2.

Quarter and core the quinces then cut into segments and place in a roasting tray so that they fit snugly in a single layer. Pour over the syrup, then tightly cover the tray with foil. Bake for 30–45 minutes until softened slightly but still retaining some resistance. Remove the foil and return to the oven for 15 minutes until the quince has softened further and the syrup has reduced slightly. Set aside to cool a little until just warm.

Meanwhile, loosely swirl the honey through the mascarpone and chill until serving.

Serve spoonfuls of quince and mascarpone with the meringue, pouring a little of the quince roasting syrup on top.

Lemon pollen pie

BRAMBLE HONEY

There is absolutely nothing better than a lemon meringue pie. Pollen in the shortcrust gives leopard spots of earthy sweet crunch.

To make the pastry, mix the flour, sugar, bee pollen and salt in a large bowl. Grate over the cold butter and rub together quickly to a breadcrumb consistency. If the butter becomes soft or starts to melt, pop the bowl in the fridge for 10 minutes before continuing. Bring the dough together into a ball, squash the top down to flatten slightly, wrap with cling film (plastic wrap) and chill for 30 minutes.

Roll the chilled pastry out to 3mm (⅛in) thick and large enough to line a 23cm (9in) tart tin (pan) with excess, then carefully drape into the tin and trim off any excess around the edge beyond 3cm (1¼in). Chill in the fridge for at least 30 minutes, even overnight.

Preheat the oven to 200°C/400°F/gas mark 6.

Remove the tart case from the fridge. Cut a square of baking parchment large enough to line the pastry with excess. Scrunch into a ball then pull out flat; the softened crinkles will make it much easier to gently line your pastry. Place the crumpled sheet in the tart case, and fill with baking beans (pie weights), carefully rolling them to the edges. Bake for 15–20 minutes, then remove the beans and parchment and bake for another 15–20 minutes until very lightly golden. Allow to cool.

Make the filling. Use some of the measured water to mix the cornflour into a paste in a small bowl. Pour the remaining water into a small saucepan with the lemon zest and sugar. Bring to the boil, then remove from the heat, add the cornflour paste and stir thoroughly.

Bring to a simmer, stirring constantly for a minute or two as it thickens. Remove from the heat and vigorously stir through, one at a time, the honey, egg yolks, lemon juice and butter. If it loses its thickness, return to a moderate heat while stirring for a minute or so.

Remove from the heat and spoon into the pastry case. Allow to cool completely before making the meringue.

Preheat the oven to 160°C/320°F/gas mark 2.

Plop the egg whites into a large mixing bowl and whisk until bubbly but not foamy. Add the honey and whisk thoroughly before adding the sugar 2 tablespoons at a time. Once the egg whites are glossy and have formed stiff peaks, spoon over the tart.

Bake for 15–20 minutes until golden, and allow to cool before serving.

Makes one 23cm (9in) tart

For the pastry

250g (9oz) plain (all-purpose) flour

75g (2½oz) caster (superfine) sugar

2 tbsp bee pollen

½ tsp salt

150g (5oz) cold butter

For the filling

100ml (3½fl oz) water

25g (1oz) cornflour (cornstarch)

Grated zest and juice of 2 or 3 small lemons (about 150ml/5fl oz of juice)

125g (4½oz) caster (superfine) sugar

2 tbsp honey

4 large egg yolks (save the whites for the meringue)

45g (1½oz) salted butter

For the meringue

4 large egg whites

½ tbsp bramble honey

240g (8½oz) caster (superfine) sugar

Michelle Polzine's 10 layer honey cake

WILDFLOWER HONEY

This stupendous ten-layered honey cake is one of only a couple of times I will permit burning honey on purpose (the other being on page 219). Medovik is a honey cake from Russia and Ukraine which is characterised by multiple wafer-thin layers of honeyed sponge, interlaced with plenty of sour cream or condensed milk. Michelle Polzine decided to double its height, burn the honey and use double (heavy) cream and dulce de leche, creating a world-famous cake synonymous with her much loved and much missed 20th Century Café in San Francisco.

This really is the cake to end all cakes. It stops everyone in their tracks, and despite its vertiginous size you won't have any leftovers; just the right balance of sweet, thanks to that burnt honey. Here is Michelle's original recipe, followed by a coffee and toasted walnut version, from me. I have kept the odd Americanism here – 'frosting' feels so much more appropriate to thick pillowy double cream than icing.

Preheat the oven to 190°C/375°F/gas mark 5.

Use a dark marker pen to trace 11 circles 23cm (9in) in diameter onto baking-sheet-sized sheets of baking parchment, using the base of a springform cake tin (pan) to trace around.

For the burnt honey, bring the honey to a simmer in a medium saucepan over a medium heat. After a few minutes, it will begin to foam intensely. Stir it occasionally with a wooden spoon and pay close attention: as soon it starts to smoke, reduce the heat to low and cook for another 30 seconds. Remove from the heat and swirl the honey for a minute to release some heat, then set the pan down and add the water (be careful; it will steam and sizzle!). Once it stops bubbling, give it a stir and pour into a heatproof measuring jug (cup). Stir in enough hot water to make 200ml (7fl oz). (The burnt honey can be made ahead; stored at room temperature, it will keep indefinitely.)

For the cake, in a medium heatproof bowl, combine the honey, burnt honey, sugar and butter, and set the bowl over a pan of simmering water. Crack the eggs into a separate bowl and set aside. In a small bowl, combine the bicarbonate of soda, salt and cinnamon.

Whisk the honey mixture. When the butter has melted and the mixture is hot to the touch (but not so hot it will burn you), add the eggs all at once, whisking.

Continued

Serves 16–20

For the cake

150ml (5fl oz) wildflower (or other mild) honey

60ml (2fl oz) burnt honey (see below)

165g (5¾oz) caster (superfine) sugar

170g (6oz) cold unsalted butter, cubed

5 large eggs

1¾ tsp bicarbonate of soda (baking soda)

1¼ tsp rock or coarse (kosher) salt

1 tsp ground cinnamon

1 tbsp cold water

360g (12½oz) plain (all-purpose) flour

For the burnt honey

200ml (7fl oz) wildflower (or other mild) honey

2 tbsp water, plus extra as needed

For the frosting

120ml (4fl oz) burnt honey (see above)

1 x 400g (14oz) can of dulce de leche

1 tsp fine sea salt

1.5 litres (3¼ pints) double (heavy) cream

Whisk until the mixture becomes hot again, then whisk in the bicarb mixture. The batter will begin to foam and smell a little weird; that's normal. Remove from the heat, whisk in the cold water, and let cool until warm but not hot, then sift in the flour and whisk until completely smooth.

Place a piece of baking parchment tracing-side-down on a baking sheet. Use an ice-cream scoop or measuring jug to spoon about 90g (3¼oz) of the batter onto the circle, and use an offset spatula to evenly spread it to the circle's edges. It will seem like just barely enough batter. If you have another baking sheet, prepare a second layer. Bake both layers, two at a time, for 6–7 minutes, rotating the sheets halfway through, until they turn a deep caramel colour and spring back to the touch. Do not overbake! Repeat with the remaining layers, reducing the bake time by a minute or two if reusing hot baking sheets. When each layer is done, slide the baking parchment off the baking sheet. Peel the cake layers off the parchment while they are still warm, but do not stack them until they're completely cool.

Reduce the oven temperature to 120°C/250°F/gas mark ½ and allow the cake layers to cool for 30 minutes. Choose your least favourite layer and return it to the oven on a parchment-lined baking sheet. Bake until it turns a deep reddish-brown and becomes very dry, about 15 minutes. Allow to cool before transferring it a food processor and grinding it to fine crumbs. Set aside.

For the frosting, in a medium bowl, whisk the burnt honey, dulce de leche and salt. Slowly pour in 180ml (6fl oz) of the cream, and whisk until smooth. Refrigerate until completely cooled, about 30 minutes.

In a stand mixer with a whisk attachment, whip the remaining cream to soft peaks, about 6 minutes on a medium speed. Add the cooled honey mixture and whip to medium-stiff peaks. If your mixer holds less than 2 litres (4 pints), make the frosting in two batches and combine in a large bowl. Refrigerate the frosting while you assemble the cake.

To assemble the cake, place your first layer on a 25cm (10in) cardboard cake circle or a flat serving plate. Spoon 150–160g (5½–5¾oz) of frosting on top and use an offset spatula to spread the frosting evenly to the edges. Top with the second cake layer and repeat the frosting and stacking process with all 10 layers. Don't be afraid to manoeuvre the cake to align the cake layers as you continue stacking. (If you have a bench scraper and cake turntable, hold the scraper at a right angle from the cake and turn the cake to smooth and straighten it as you go.) After your tenth layer, spread a final layer of frosting on the top and use any leftover frosting to smooth out the side of the cake. Gently press the reserved cake crumbs on the sides of the cakes until completely coated. Scatter some more crumbs on top, as you wish.

Refrigerate the cake overnight and serve chilled. The cake can be made up to 2 days in advance; refrigerate leftovers for up to 3 days.

Coffee and walnut version

I have a Californian friend called Alex who loves to make Michelle Polzine's honey cake. Alex is a coffee (and cake) nerd, and thanks to him and his hand-me-down grinder, I've watched way too many James Hoffman videos and now call a cafetière a French press and I don't even press it. This take on the classic British coffee and walnut fête cake is for Alex.

Considering I've used instant coffee in this recipe, perhaps I should give the fancy equipment back with my tail between my legs and beg forgiveness. The toasted nuts and bitter coffee complement the burnt honey and sweet dulce de leche perfectly, so hopefully it'll distract him.

Extra ingredients/swaps

200g (7oz) walnut halves

4 tbsp instant coffee, mixed with 1 tbsp boiling water and left to cool

Soft light brown sugar in place of caster (superfine) sugar

METHOD ADAPTATIONS

Toast the walnut halves and chop half very finely and half a little more roughly.

Mix half the cooled coffee into the cake batter with the finely chopped walnuts, and beat the other half into the frosting mixture.

Mix the roughly chopped walnuts into the blitzed sponge dust for decorating.

Fig leaf panna cotta

HEATHER HONEY

'The perfect dessert doesn't exis...' Yes it does. If you haven't made panna cotta before, you really should. Plus, I defy you not to giggle at the jiggle. Simple to make and utterly delicious, you can proudly deliver one of these wobbly wonders back over the neighbour's fence in return for pinching a few leaves off their fig tree. If you can't get hold of any fig leaves, roast a couple of figs (see page 220), and use those to infuse the cream instead.

Makes 6 x 150ml (5fl oz) individual puddings using moulds

4 fig leaves

200ml (7fl oz) full fat milk

700ml (1½ pints) double (heavy) cream

45g (1½oz) caster (superfine) sugar

1 large strip of pared lemon zest

3 gelatine leaves (each about 7.5cm/3in)

1 tbsp heather honey

Preheat the oven to 160°C/325°F/gas mark 3. Place the fig leaves directly on the oven rack and toast until fragrant and dry – about 5–10 minutes should do it but keep an eye on them.

Heat the milk, cream and sugar gently in a heavy-based saucepan set over a low heat. When the fig leaves are ready, crumble them into the pan, add the lemon zest, and keep warming gently for 5 minutes (do not allow to simmer).

Remove the pan from the heat, cover, and leave to cool, stirring occasionally, for 1½–2 hours. Alternatively, you can store this in the fridge overnight to allow the flavours to infuse more deeply.

Soak the gelatine leaves in water until soft.

Place the panna cotta moulds in a roasting tin.

Strain the cream through a fine sieve and/or muslin (cheesecloth) set over a clean saucepan. Discard the zest and fig leaf shards. Return the cream to the heat, bring to the boil then immediately turn off the heat. Thoroughly squeeze out the water from the soft gelatine leaves and whisk them into the cream. Stir through the honey, then carefully pour the mixture into the moulds. Allow to cool, then refrigerate overnight to set.

To loosen the panna cotta from their moulds, briefly dip the moulds in hot water before inverting onto plates. Serve with fresh or roasted figs (see page 220).

CHAPTER 5

Ices

215 Honey on toast ice cream

216 Roasted black sesame ricotta ice cream

219 Burnt honey dulce ice cream

220 Honey roast fig and Amaretto ice cream

222 Three granitas

Honey on toast ice cream

CLOVER HONEY

Honey on toast is right at any time of day or night. It doesn't need infusing and freezing into this delicious, comforting ice cream to pass at 10pm on a Thursday when you're having a difficult week, but sometimes only ice cream will cheer you up. Make this for those times.

I once had (untoasted) brown bread ice cream, half chewy with flecks of blitzed, malty wholemeal. It tasted a little like if you froze a bowl of soggy Weetabix, only creamier, and was kind of incredible, but that's not the sort of bread to bring much pleasure at any time of day. For this ice cream, I suggest a good crusty slice or end of white or wholegrain bread to get all the toasty caramelised flavour from its Maillard-ed outer. If you want to use sourdough, go for a mild one.

Bread ice cream is indeed 'a thing', from at least the 17th century in England and Ireland, then popping up in Jane Grigson's English Food *as an ice cream that works without churning, even with the traditional custard base. Fergus Henderson makes a delicious one with Armagnac.*

Clover honey is my favourite honey to spread on toast, so I've suggested it for this recipe. As the honey isn't heated, go for a good-quality raw one to enjoy all of its flavour.

Makes 600ml (1¼ pints)

1 slice of good crusty bread

300ml (10fl oz) double (heavy) cream

4 eggs, separated, yolks lightly beaten

100g (3½oz) clover honey

Preheat the oven to 180°C/350°F/gas mark 4.

Toast the bread then cut into small pieces.

Warm the cream gently in a saucepan, remove from the heat and stir through the toast, reserving a few pieces for garnish. Steep for 30 minutes, then strain through a sieve, pressing the soft toast to release as much cream as possible. Allow the cream to cool fully.

In one bowl, whisk the egg whites to stiff peaks, then whisk in the honey. In another bowl, whip the cream to soft peaks, then add the egg yolks and briefly whip through. Fold the cream mixture thoroughly into the egg whites, transfer to a lidded freezer container and freeze for at least 4 hours.

Serve with the reserved toast crumbled over the top and an extra drizzle of honey.

Roasted black sesame ricotta ice cream

BRAMBLE HONEY

The smell of roasted black sesame crushed in a pestle and mortar is utterly marvellous. Even if you have zero intention of making ice cream (although this one is beautiful and very simple, so please do) you could just lazily pummel some sesame for a couple of minutes for the sheer pleasure of it. Your olfactory wellbeing will thank you. The proportions of sesame, honey and zingy cut-throughs are all up for grabs here; change to suit your mood. Pairs well with slowly roasted plums.

Makes 1.5 litres (3¼ pints)

Pinch of salt

250g (9oz) ricotta

4 tsp black sesame seeds (untoasted is fine)

80g (3oz) bramble honey

Grated zest of ½ lime

300g (10½oz) double (heavy) cream

½ tsp vanilla extract

3 large eggs, separated

125g (4½oz) caster (superfine) sugar

½ tsp pomegranate molasses

Stir the salt through the ricotta, then suspend in muslin (cheesecloth) over a bowl for a few hours, giving it the odd squeeze and twist, to remove some water.

If your sesame seeds did not come already toasted, do so now. Toss lazily in a pan on the hob (stovetop) until fragrant; it won't take long. You can also do this on a tray in the oven, but it is slower and harder to tell when they are done.

Crush the sesame seeds in a pestle and mortar until finely ground. Scrape out with a teaspoon into a small bowl and mix with 40g (1½oz) of the honey. I advise against adding the honey to the pestle and mortar; sticky chaos reigns.

In another bowl, add the lime zest to the cream, dribble in the vanilla, then whip until very soft peaks form. Don't over-whip until stiff or it will be difficult to incorporate with the rest of the ingredients and you will end up with vapid pockets of plain frozen cream. If you think you haven't whipped enough, you've whipped plenty.

In a separate bowl, whisk the egg whites and sugar to soft peaks.

Combine the remaining honey, the egg yolks and strained ricotta until mixed; a slight uneven result is just fine. Fold in the whipped cream, then the whipped egg whites and sugar.

Drizzle the black sesame honey and pomegranate molasses on top and swirl through very briefly to marble. Gently transfer to a lidded freezer container and freeze for at least 6 hours.

Burnt honey dulce ice cream

BURNT WILDFLOWER AND BORAGE HONEY

This heaven-in-a-spoon is the second time only I will permit burning honey on purpose, and is the love child of Michelle Polzine's 10 layer honey cake on page 207. The bitterness of the burnt honey works perfectly in creamy, cold ice cream. You could just make extra of the cake's frosting, then whisk some egg whites and fold in with the yolks to make this, but if you're not making the cake, follow the quantities here. I like to swirl some of the dark burnt honey and toffee-coloured dulce de leche through the ice cream for added drama. Sometimes I also like to swirl through cacao nibs for a little fruity toasty crunch.

Whisk the egg whites with the sugar into soft peaks. In another bowl, whip the cream and condensed milk to soft peaks. Stir the burnt honey, runny honey and dulce de leche into the egg yolks. Fold the yolk mixture into the whipped cream, then fold in the whisked egg whites.

Warm the 25g (1oz) dulce de leche slightly so that it is a little runnier. Swirl the 25g (1oz) burnt honey and the warmed dulce through the ice cream, transfer to a lidded freezer container and freeze for at least 4 hours.

Makes 1.5 litres (3¼ pints)

4 large eggs, separated

100g (3½oz) caster (superfine) sugar

350ml (12fl oz) double (heavy) cream

50ml (2fl oz) condensed milk

50g (2oz) burnt honey (see page 207), plus 25g (1oz) for swirling

50g (2oz) runny borage honey

75g (2½oz) dulce de leche, plus 25g (1oz) for swirling

Honey roast fig and Amaretto ice cream

APPLE BLOSSOM HONEY

Honey roast fig, but also fig roast honey, as the honey drizzled over the figs before baking takes on all their fruity sappy flavour as they gently roast. Leaving some of the figs chopped creates little soft jammy chunks, which together with the biscuity crumbs of the amaretti and little sip of Disaronno make for a wicked summer sundae.

There are two main schools of thought for no-churn ice cream. In the Mary Berry camp is whisked egg whites, in the Martha Stewart camp is condensed milk. I find the egg whites create an easily scoopable ice cream, while the condensed milk approach is very creamy but difficult to scoop. Both are of course delicious, so this recipe is a happy compromise between the two for creamy scoopability, to get you eating your dessert as soon as possible...

Makes 1.5 litres (3¼ pints)

8 figs

100g (3½oz) apple blossom honey

50g (2oz) amaretti biscuits (cookies), plus extra to serve

4 large eggs, separated

100g (3½oz) caster (superfine) sugar

350ml (12fl oz) double (heavy) cream

50ml (2fl oz) condensed milk

4 tbsp Disaronno

Preheat the oven to 160°C/320°F/gas mark 2.

Cut the figs into quarters, place on a baking tray in a single layer, drizzle with half the honey and roast for about 10 minutes, keeping an eye on them and removing before they or the honey turn dark; they should be just caramelised around the edges. Set aside to cool.

Meanwhile, blitz or bash the biscuits into a fine crumb and set aside. Whisk the egg whites with the sugar to soft peaks, and in another bowl, whip the cream with the condensed milk to soft peaks. Stir together the remaining honey, the egg yolks and Disaronno.

Save the honey from the fig roasting tray. Chop half the cooled figs into large chunks, and pulp the remaining half using the side of a knife. Add the figs and fig honey to the yolk mixture.

Fold the yolk mixture into the whipped cream, then fold in the whisked egg whites. Gently swirl through the amaretti biscuits to create a marbled effect. Transfer to a lidded freezer container and freeze for at least 6 hours.

Serve with extra amaretti biscuits crumbled over the top, if you like.

Three granitas

Granitas originate in Sicily, and these chunky ones are in the flakier Palermo style, which requires just a fork and a freezer to achieve. So retro, so chic, and so good at smuggling a cocktail into a dessert. Two of these recipes showcase particularly characterful honeys, which enjoy a robust kick from the spirits and a sublime palate mellowing from freezing. Feel free to make these sans alcohol, but it's an utter delight to enjoy as a stiff palate cleanser after dinner on a summer's evening. My partner and I like to eat a little from frosted coupes, popped in the freezer for a minute while we do the washing up.

Eucalyptus honey granita

Mix all the ingredients together, pour into a shallow, sealed container and place in the freezer for 2 hours, then gently flake the ice crystals with a fork and return to the freezer for another hour. Repeat hourly until loose but fully frozen.

Makes 500ml (1 pint)

250ml (8½fl oz) water
100ml (3½fl oz) eucalyptus honey
Juice of 1 lemon
Juice of 1 lime
1 tsp bitters
Double shot of vodka

Mezcal blood orange hibiscus

Simmer the hibiscus flowers in the water until vivid pink. Remove the flowers, then add the sugar. Dissolve and reduce slightly for 5–10 minutes over a low–medium heat. Leave to cool, then stir together with the remaining ingredients.

Pour into a shallow, sealed container and place in the freezer, then gently flake the ice crystals with a fork and return to the freezer for another hour. Repeat hourly until loose but fully frozen.

Makes 500ml (1 pint)

1 tbsp hibiscus flowers
250ml (8½fl oz) water
25g (1oz) sugar
75ml (2½fl oz) Yucatan honey
Juice of 2 blood oranges
3 shots of mezcal

Campari watermelon mint

Blitz the watermelon and mint in a blender then leave to infuse in the fridge for 1–2 hours.

Strain through a piece of muslin (cheesecloth), then mix with the remaining ingredients and pour into a shallow, sealed container. Place in the freezer for 2 hours, then gently flake the ice crystals with a fork and return to the freezer for another hour. Repeat hourly until loose but fully frozen.

Makes 900ml (scant 2 pints)

Flesh of ½ small watermelon (500g/1lb 2oz)

Small sprig of mint, stems removed

250ml (8½fl oz) water

50ml (2fl oz) acacia honey

2½ shots of Campari

3 shots of gin

Squeeze of lime juice (up to 1 lime)

CHAPTER 6

Drinks

228	Fiery ginger kombucha
231	No time to tepache
232	Tepache Bourbon sour
235	Bergamot Bee's Knees
236	Stormy black lime
238	Smoked mezcal Margarita
241	Sesame Old Fashioned

Fiery ginger kombucha

RAW WILDFLOWER HONEY

Kombucha is sweet tea that you forget about for a week or so, then bottle so that it becomes enticingly fizzy. Meanwhile, it ferments. That's it. I explain it as such, because many people associate ferments with being faffy, strange, a bit worthy and probably-not-actually-nice, rather than easy, delicious and just happening to be good for you. This one is the latter.

Raw honey is a great flavourful alternative to the sugar usually used in kombucha, pre-packed with its own beneficial bacteria and yeasts to get things going. I've used black tea and rooibos, but you can use green tea too. You can buy your 'scoby' (Symbiotic Culture of Bacteria and Yeast) online, a little jelly pancake that you plop into the sweet tea to get things going, and then watch with morbid fascination as another jelly pancake grows underneath it, and then you can be that sort of person who gives homemade scobys to friends to start their own kombuchas. I must confess, as I write this, I have left my sweet tea on the side for a month, not the recommended 7–10 days, so it will definitively not be sweet but very vinegary. But then that's a joy to cook with too, and now I have a splendid new pancake to re-make a sweeter one, and I can give my old pancake to a friend, so really you can't go wrong here.

Makes 1.5 litres (3¼ pints)

1.75 litres (3½ pints) water
2 organic black teabags
2 organic rooibos teabags
150g (5oz) raw wildflower honey
1 small scoby pancake
1 large piece of organic ginger (7.5–10cm/3–4in long)

Boil the water in a saucepan then add all the teabags and brew for 10 minutes. Remove the teabags and, once cooled, stir through the honey. Pour into a 2-litre (4-pint) glass jar with a wide mouth for brewing. Add the scoby pancake and the little liquid it comes with. Cover the top of the jar with muslin (cheesecloth) and store at room temperature out of direct sunlight for 7–10 days.

After 7 days, taste a little. The sugars in the honey will slowly be 'eaten' during the fermentation, so it will become less sweet and more acidic. If still sweet and very tea-like but not particularly interesting, leave for a day and taste again. Continue until you're interested but not put off.

You can drink the kombucha at this stage, after its first ferment. To introduce additional flavours and fizz, we do a short second ferment.

Gently wash the ginger in water then grate, skin on, saving the juice and pulp. Add 2 tablespoons to each of 3 x 500ml (1-pint) bottles (I use swing-top). Fill the bottles with the fermented tea (saving the scoby and a little liquid for your next ferment), leaving a little space at the top. Leave to ferment at room temperature with the lids on for 1–3 days until fizzy; you can check every day – it'll re-fizz after capping.

Once fizzy, it's ready and should be stored in the fridge. It will keep for a couple of months, but I normally drink it within a fortnight. Serve chilled.

No time to tepache

RAW AND YUCATAN HONEY

THIS drink. It's Mexican, it's fermented, it's perfect with honey, it uses fruit scraps that would have gone in the bin, it makes amazing cocktails (see page 232) and it's endlessly variable. Need I say more? No, but I will. It is traditionally made using an unrefined cane sugar called piloncillo, and pineapple skins; their natural yeasts kick-start the fermentation process to create a zingy, sweet, spiced drink. Using raw honey with its own yeasts and full flavour is a great alternative to piloncillo, which is a little elusive here in London. My fridge is full of sticky jars of forgotten tepaches; anything I leave fermenting in the kitchen for 'a day or so' will stay there for eternity. After my fourth forgotten tepache, I started muttering 'clearly no time to tepache' while shuffling more sticky jars into the fridge. Luckily they're all delicious in their own way – I've given you two variations here. I like to call them my cuvées, my exclusive vintages. Don't be like me; set an alarm on your phone to check your tepache in 2 days!

Each makes 500ml (1 pint)

For peach and peppercorn

1 x 8cm (3in) cinnamon stick

4 cloves

1 whole star anise

4 allspice berries

½ tsp pink peppercorns

100ml (3½fl oz) raw honey

100ml (3½fl oz) Yucatan honey

250ml (8½fl oz) warm water

4 organic peaches

For apricot and caraway

1 x 8cm (3in) cinnamon stick

4 cloves

1 whole star anise

2 allspice berries

1 tbsp caraway seeds

200ml (7fl oz) raw honey

250ml (8½fl oz) warm water

6 organic apricots

Toast the spices in a hot, dry pan until fragrant, then leave to cool.

Stir together the honey(s) and 100ml (3½fl oz) of the warm water.

Cut the peaches or apricots in half, and pop the halves inside a 1-litre (2-pint) jar. Add the toasted spices, then pour over the honey water. Top up with the remaining water as necessary, to cover the fruit. Weigh the fruit down under the liquid using a fermentation weight or something heavy and inert like a pestle, cover the opening with muslin (cheesecloth) and leave on the side out of direct sunlight to ferment for 2–4 days, until frothing on top, then strain and store in the fridge. It will keep for a couple of months, but I normally drink it within a fortnight.

Serve watered down over ice with fizzy or still water, or use in cocktails. Leave it fermenting for longer if you'd like tepache vinegar. The leftover peach or apricot halves are delicious served with cheese and cured meats.

Tepache Bourbon sour

My next book will be called 101 Ways to Use Honey Tepache.

Chill a whisky glass in the freezer while you toast the cinnamon stick in a dry, moderately hot pan until fragrant. Allow to cool.

Shake the wet ingredients with plenty of ice and strain into an ice-filled glass. Decorate with the cinnamon stick.

Serves 1

1 cinnamon stick
50ml (2fl oz) Bourbon
20ml (4 tsp) lemon juice
25ml (1fl oz) honey tepache of your choice (see page 231)
3 dashes of Angostura bitters
15ml (1 tbsp) egg white

Bergamot Bee's Knees

BORAGE HONEY

I love cocktails, I love bees, I love citrus, and I love a good story. The Bee's Knees is not just a great turn of phrase, along with the Cat's Pyjamas, it's also a cocktail, thought to originate in prohibition-era New York, where honey and lemon were used to take the edge off bathtub gin. The 'official' recipe now has equal parts lemon and orange juice, for a slightly more quaffable drink.

Using seasonal citrus varieties is a great way of adding a fresh twist to classics when you feel like it; one weekend in November I picked up some bergamots, courtesy of Natoora, which have a spicy violet perfume that brings some sultry sophistication to this simple drink. Use carefully; bergamot juice is face-meltingly sour, so don't be tempted to sub out all the lemon or orange. The magic is in its perfumed zest, so I've used it three ways here: a rough strip to rim the glass and add to the shaker, a classic twist (thin piece rolled round a chopstick to create a corkscrew) to garnish, and a round disc piece, called a cheek, to squeeze over a flame for the final party trick, releasing the bergamot oil over the drink.

You can riff off from this perfectly delicious baseline in several ways: swap the gin for rum and you get a Honeysuckle. Use blood orange juice and top up with cava for a sunset-coloured spritz, and up the floral violet note with some lavender honey.

Serves 1

5ml (1 tsp) hot water
10ml (2 tsp) borage honey
Bergamot peel: 1 rough strip, 1 classic twist, 1 'cheek' disc
50ml (2fl oz) gin (2 shots)
25ml (1fl oz) lemon juice
15ml (1 tbsp) orange juice
10ml (2 tsp) bergamot juice

Chill a small coupe in the freezer. Stir the hot water into the honey until mixed. Remove the coupe from the freezer and rim with the rough strip of peel. Combine the wet ingredients with the peel strip in a shaker with ice, shake until very cool. Pour into the chilled coupe and garnish with the twist.

For the flaming cheek trick, take your bergamot rind cheek, aka a disc of skin, in one hand, holding it with a thumb on one edge and forefinger on the top edge, rind facing your drink, about 10cm (4in) above the surface. In your other hand, take a lighter and light and gently warm the rind for a couple of seconds. Then, eyebrows and long hair safely kept back, hold the flame just away from the rind and pinch the rind quickly between your finger and thumb so that it bends in half towards the flame. The essential oils will burst over your drink and set light, causing a quick flash!

Stormy black lime

BUCKWHEAT HONEY

For a deeper flavour from the gloriously fusty mummified limes, infuse the rum with the black lime slices overnight, a week, or longer.

Chill a highball glass in the freezer. Using a small, very sharp serrated knife, cut the black lime into 4 slices. Make the 1:1 honey syrup by mixing 1 teaspoon of honey with 1 teaspoon of hot water. Remove the glass from the freezer and fill with ice.

Shake the lime slices, fresh ginger, rum, lime juice, honey syrup and bitters with ice, and strain over the ice-filled glass. Transfer a slice of black lime and the ginger into the glass. Top with the ginger beer and briefly stir.

Serves 1

1 black lime

10ml (2 tsp) 1:1 buckwheat honey syrup (see method)

1 slice of fresh ginger

50ml (2fl oz) dark spiced rum

20ml (4 tsp) lime juice

4 dashes of Angostura bitters

90ml (3fl oz) fiery, good-quality ginger beer (not too sweet)

Smoked mezcal margarita

SMOKED HONEY

You can use smoked limes OR smoked honey, or just a smoky mezcal, but I'm a bit of a sucker for a strong, punchy drink, so sometimes I go for all three. If smoking the limes and honey yourself, be very gentle and taste as you go; too long and they'll taste like a pub carpet. Pair with tacos (see pages 100 and 103) or the smoked lime and honey chicken on page 126.

Serves 1

1 tbsp lime juice (smoked if you like, from 1 lime; see method)

15ml (1 tbsp) honey (smoked if you like; see method)

1 tbsp Tajin seasoning

1 fine slice of red chilli

60ml (2fl oz) mezcal

Chill a whisky glass in the freezer while you smoke the limes and honey, if doing so.

To smoke, light a handful of good lumpwood charcoal on the BBQ (grill), to one side. Once hot, add a small chunk of smoking wood such as cherry, or soaked chips, and place the grill rack on top. Add a halved lime, and the honey in a shallow dish, to the grill rack, furthest away from the smoking wood and charcoal. Pop the lid on, have the vents open slightly, and smoke for about 20 minutes.

Spread the Tajin on a small plate. Remove the glass from the freezer and wet the rim with half a smoked lime. Lightly dip the rim in the Tajin and fill with ice and the chilli slice. Add the mezcal, lime juice and honey to a shaker with ice and shake vigorously. Strain and pour.

Sesame Old Fashioned

SMOKED, CHESTNUT OR OAK HONEY

As you can tell from the proliferation of sesame through this book, I think sesame and honey are a wonderful pairing. The technique for smuggling sesame into an Old Fashioned comes from Sicilian sommelier Santo Borzi of Bar 190 at The Gore Hotel in London, and uses a technique called 'fat-washing', fat being the sesame oil, the outcome of which tastes much nicer than it sounds. Borzi uses actual smoke under a cloche when presenting his masterpiece; here I've replaced the classic sugar syrup with smoked honey. I recommend using the leftover whiskyed sesame oil with soy sauce, a little ginger and rice vinegar, as a coating for garlic-honey roasted chicken.

First infuse (fat-wash) the whisky: pour into a large glass jar. In a small pan, gently heat the sesame oil until just above blood temperature (around 37°C/98°F – toasty but not hot), then pour into the whisky jar and stir. Leave to infuse for 6 hours, stirring gently every hour, then seal and put in the freezer overnight – the oil will rise to the top and solidify during this time.

Scoop out and discard the frozen oil, then pass the infused whisky through 2 layers of coffee filters or muslin (cheesecloth), to filter out any remaining oil residue (and repeat if need be). Pour the whisky into a clean jar or bottle, seal and store at room temperature; it will keep for a good 2 weeks.

To make the cocktail, add the whisky, bitters and honey to a shaker or large glass filled with ice, and stir for a minute or two. Place a large ice cube in a short tumbler glass and strain the drink over. Decorate with the orange twist.

Serves 1

2 shots of sesame-infused whisky (see below)

3 dashes of Angostura bitters

1 tbsp honey (for smoked honey, see page 238), loosened with a drop of hot water

1 twist of orange peel, to decorate

For the sesame-infused whisky

250ml (8½fl oz) whisky or Bourbon

60ml (2fl oz) toasted sesame oil

Index

A
Acacia honey 25, 26, 30, 39
  Campari watermelon mint granita 223
  charred radicchio and pickled carrot salad with sesame-seared tuna 88
  cut glass paprenjacs 171
  honeycomb bread 160–2
achiote orange tacos with jackfruit and cauliflower 100
almonds: almond briouats 181–3
  almond, olive and raspberry loaf cake 184
  almond, walnut and stem ginger baklava with orange blossom honey syrup 175
  blood orange, almond and dark chocolate babka 157
  griddled green mustard salad with smoked almonds 86
  halwa chebakia 178–80
Amaretto: honey roast fig and Amaretto ice cream 220
Apis cerana 39
  A. mellifera 39, 41
apple blossom honey 39, 42
  honey roast fig and Amaretto ice cream 220
  honey tarte tatin 191–3
  rose roast quince with hazelnut meringue and honeyed mascarpone 203
apples: honey tarte tatin 191–3
apricots: apricot and fennel croissant swirls 144–9
  apricot caraway tepache 231
  gochujang apricot sticky wings 99
  saffron and apricot honey buns 154
aubergines (eggplants): dark beef and aubergine curry 129
autumn 20
avocado: honeyed chipotle lamb tacos 103

B
babkas 156–7
baklava 174–5
Basque cheesecake, cardamom 200
basswood honey 40
beans, broken 96
bee pollen: lemon pollen pie 204
  pollen and pistachio baklava with honeyed hibiscus syrup 174
beef: dark beef and aubergine curry 129
beekeeper's year 10–23
bees: drone bees 15, 20
  eggs 11
  how bees find flowers 48
  pollination 45–9
  queen bees 11, 12, 15, 16, 18, 20
  species of 48
  worker bees 11, 12, 15, 16, 33, 35, 45, 46, 48
beeswax 11, 16, 20, 23
beetroot (beets): harissa hazelnut beetroot with raw cavolo and dilly red peppers 109
bergamot bee's knees 235
birch honey: home-smoked honey 65
biryani: goat biryani with jackfruit and lime leaf rice 130
black garlic and lime tomatoes 85
black lime, stormy 236
blood orange: almond and dark chocolate babka 157
bloom calendars 54
borage honey 39

bergamot bee's knees 235
black garlic and lime tomatoes 85
burnt honey dulce ice cream 219
borders, maximising 52
Bourbon: tepache Bourbon sour 30, 232
bramble honey 39
  lemon pollen pie 204
  roasted black sesame ricotta ice cream 216
bread: broken beans 96
  garlic fermented honey and miso sourdough 140–3
  Helen's foolproof flatbreads 132–3
  honey on toast ice cream 215
  honeycomb bread 160–2
  preserved lemon focaccia 152–3
briouats, almond 181–3
broken beans 96
buckwheat honey 26, 30, 39
  honey hong shao rou 136
  stormy black lime 236
bulbs 52, 57
bulgur wheat: minted lamb chops with herbed grains, pomegranate and sherry vinegar 113–15
bumblebees 52, 56, 57
buns, saffron and apricot honey 154
burnt honey dulce ice cream 219
burrata: fennel roast squash with burrata, pickled mushrooms and crispy onions 106
buttermilk: harissa buttermilk dressing 72

C

cabbage: goat biryani with jackfruit and lime leaf rice 130
  griddled green mustard salad with smoked almonds 86
  lime leaf slaw 122
cakes: almond, olive and raspberry loaf cake 184
  coffee and walnut cake 209
  honey nut corn cake 197
  loaf cakes 194–5
  Michelle Polzine's 10-layer honey cake 207–8
  orange, yogurt and cardamom loaf cake 185
Campari watermelon mint granita 223
caramel: burnt honey dulce ice cream 219
  Michelle Polzine's 10-layer honey cake 207–8
caraway seeds: apricot caraway tepache 231
cardamom: cardamom Basque cheesecake 200
  cardamom oats 78
  orange, yogurt and cardamom loaf cake 185
  pistachio, rose and cardamom babka 156–7
carlin peas: paprika parched peas with butter-roast fennel and garlic green peppercorn mash 110–12
carrots: charred radicchio and pickled carrot salad with sesame-seared tuna 88
  fermented fennel kimchi 68–9
  ginger roasted carrots with chilli and chives 82
cauliflower, achiote orange tacos with jackfruit and 100

cavolo nero: harissa hazelnut beetroot with raw cavolo and dilly red peppers 109
chamomile 52
chebakia, halwa 178–80
cheesecake, cardamom Basque 200
Chelsea chop 57
chemicals 54, 57
cherry honey: home-smoked honey 65
chestnut flour: brown butter / oak / chestnut madeleines 163–5
chestnut honey 25, 40
  sesame Old Fashioned 241
chicken: gochujang apricot sticky wings 99
  smoked lime and honey chicken 126
chillies: ginger roasted carrots with chilli and chives 82
  gochujang apricot sticky wings 99
  guajillo garlic prawns 92
  sour cherry chipotle hot sauce 73
Chinese (napa) cabbage: fermented fennel kimchi 68–9
chipotle chillies: honeyed chipotle lamb tacos 103
  sour cherry chipotle hot sauce 73
chives, ginger roasted carrots with chilli and 82
chocolate: blood orange, almond and dark chocolate babka 157
citrus juice: achiote orange tacos with jackfruit and cauliflower 100
climbing plants 54
clover 42–3, 52

clover honey 26, 40, 43
  cardamom oats 78
  crumpets 81
  custard tart! 188–90
  honey on toast ice cream 215
  lemon curd 74
coffee and walnut cake 209
Cook, Eliza 42
corn cake, honey nut 197
corn cobs: smoked lime and honey chicken 126
couscous: minted lamb chops with herbed grains, pomegranate and sherry vinegar 113–15
cream: burnt honey dulce ice cream 219
  cardamom Basque cheesecake 200
  custard tart! 188–90
  fig leaf panna cotta 210
  griddled peach and pistachio pavlova with lavender honey 194–6
  honey on toast ice cream 215
  honey roast fig and Amaretto ice cream 220
  Michelle Polzine's 10-layer honey cake 207–8
  roasted black sesame ricotta ice cream 216
cream cheese: cardamom Basque cheesecake 200
  honeycomb bread 160–2
croissant swirls, apricot and fennel 144–9
crumpets 81
crystallization 25–6
cucumber: smoked duck, smashed cucumber, slapped noodles 119–21
cupboard honey: broken beans 96

curd, lemon 74
curry: dark beef and aubergine curry 129
   goat biryani with jackfruit and lime leaf rice 130
custard tart! 188–90
cut glass paprenjacs 171

D
daikon: fermented fennel kimchi 68–9
dark beef and aubergine curry 129
dressing, harissa buttermilk 72
drinks 30–1, 226–41
   apricot caraway tepache 231
   bergamot bee's knees 235
   fiery ginger kombucha 228
   peach peppercorn tepache 231
   sesame Old Fashioned 241
   smoked mezcal margarita 238
   stormy black lime 236
   tepache Bourbon sour 30, 232
drone bees 15, 20
duck: smoked duck, smashed cucumber, slapped noodles 119–21
dulce de leche: burnt honey dulce ice cream 219
   Michelle Polzine's 10-layer honey cake 207–8

E
eggs: broken beans 96
   cardamom Basque cheesecake 200
   custard tart! 188–90
   griddled peach and pistachio pavlova with lavender honey 194–6
   lemon curd 74
   lemon pollen pie 204
Ellis, Hattie 29
Ethiopia 31
eucalyptus honey 30, 40
   almond briouats 181–3
   eucalyptus honey granita 222

F
fennel: fermented fennel kimchi 68–9
   paprika parched peas with butter-roast fennel and garlic green peppercorn mash 110–12
   radicchio pear salad with spiced fritto misto 91
fennel seeds: apricot and fennel croissant swirls 144–9
   fennel roast squash with burrata, pickled mushrooms and crispy onions 106
fermentation: fermented fennel kimchi 68–9
   fermenting honey 30–3
   honey jar ferments 62
fertilisers 43
field bean honey 40
   paprika parched peas with butter-roast fennel and garlic green peppercorn mash 110–12
fiery ginger kombucha 228
fig leaf panna cotta 210
figs: honey jar ferments 62
   honey roast fig and Amaretto ice cream 220
filo (phyllo) pastry: almond, walnut and stem ginger baklava with orange blossom honey syrup 175
   almond briouats 181–3
   pollen and pistachio baklava with honeyed hibiscus syrup 174
fish: charred radicchio and pickled carrot salad with sesame-seared tuna 88
   Saikyo salmon 116
flatbreads, Helen's foolproof 132–3
flowers 42, 49, 54
   Chelsea chop 57
   flowering times 54
   how bees find 48
   pollination 46–9
focaccia, preserved lemon 152–3

forest honey 36
French beans: griddled green mustard salad with smoked almonds 86
fritto misto, radicchio pear salad with spiced 91
fructose 25–6, 29, 33

G
gardens 42
   design and planting 52–5
   maintenance 56–7
   plant recommendations 58–9
   for pollinators 49
garlic: black garlic and lime tomatoes 85
   garlic fermented honey and miso sourdough 140–3
   guajillo garlic prawns 92
   honey jar ferments 62
   paprika parched peas with butter-roast fennel and garlic green peppercorn mash 110–12
garlic fermented honey 31
   garlic fermented honey and miso sourdough 140–3
   ginger roasted carrots with chilli and chives 82
   griddled green mustard salad with smoked almonds 86
   preserved lemon focaccia 152–3
gin: bergamot bee's knees 235
   Campari watermelon mint granita 223
ginger: almond, walnut and stem ginger baklava with orange blossom honey syrup 175
   fiery ginger kombucha 228
   ginger roasted carrots with chilli and chives 82
ginger beer: stormy black lime 236
glucose 25–6, 33
goat: goat biryani with jackfruit and lime leaf rice 130

spice cupboard goat with preserved lemons 132–3
gochujang apricot sticky wings 99
Godwin, Richard 30
gooseberries: honey jar ferments 62
Goulson, Dave 54
grains: minted lamb chops with herbed grains, pomegranate and sherry vinegar 113–15
granita: Campari watermelon mint granita 223
eucalyptus honey granita 222
mezcal blood orange hibiscus granita 222
grapefruit: radicchio pear salad with spiced fritto misto 91
Greek honey puffs 170
Greek thyme honey: loukoumades 170
guajillo garlic prawns 92

H
hard landscaping 55
harissa: harissa buttermilk dressing 72
harissa hazelnut beetroot with raw cavolo and dilly red peppers 109
harvesting honey 20
hazelnuts: harissa hazelnut beetroot with raw cavolo and dilly red peppers 109
rose roast quince with hazelnut meringue and honeyed mascarpone 203
heather honey 16, 17, 26, 36, 40
custard tart! 188–90
fig leaf panna cotta 210
saffron and apricot honey buns 154
heating honey 26
Helen's foolproof flatbreads 132–3
herbicides 43, 57
herbs: minted lamb chops with herbed grains, pomegranate and sherry vinegar 113–15

hibiscus flowers: mezcal blood orange hibiscus granita 222
pollen and pistachio baklava with honeyed hibiscus syrup 174
hives: annual quantity of honey 45
fermentation in the 31–3
overwintering 11–12, 23
temperature of 11, 25
home-smoked honey 65
gochujang apricot sticky wings 99
guajillo garlic prawns 92
silky squash pasta/soup 95
sour cherry chipotle hot sauce 73
honey: cooking with 24–9
crystallization 25–6
drinking honey 30–1
fermenting 30–1
harvesting 20
heating 26–9
history of 25, 26, 30, 31
moisture content 29
terroir 32–43
varietals 39–41
honey hong shao rou 136
honey jar ferments 62
honey nut corn cake 197
honey on toast ice cream 215
honey roast fig and Amaretto ice cream 220
honey tarte tatin 191–3
honeycomb bread 160–2
honeyed chipotle lamb tacos 103
honeyed mushrooms, 2 ways 66

I
ice cream: burnt honey dulce ice cream 219
honey on toast ice cream 215
honey roast fig and Amaretto ice cream 220
roasted black sesame ricotta ice cream 216
icing, orange blossom 163–5
ivy 55

J
jackfruit: achiote orange tacos with jackfruit and cauliflower 100
goat biryani with jackfruit and lime leaf rice 130
June gap 58

K
khaliat (al) nahal 160–2
kimchi, fermented fennel 68–9
kombucha, fiery ginger 228

L
lactic acid bacteria (LAB) 33
lamb: honeyed chipotle lamb tacos 103
minted lamb chops with herbed grains, pomegranate and sherry vinegar 113–15
lavender honey 29, 40
apricot and fennel croissant swirls 144–9
griddled peach and pistachio pavlova with lavender honey 194–6
lawns 52, 56
lemons: lemon curd 74
lemon pollen pie 204
preserved lemon focaccia 152–3
spice cupboard goat with preserved lemons 132–3
lettuce: griddled green mustard salad with smoked almonds 86
'lime' honey 40
lime leaves: double plum ribs and lime leaf slaw 122
goat biryani with jackfruit and lime leaf rice 130
limes: black garlic and lime tomatoes 85
smoked lime and honey chicken 126
smoked mezcal margarita 238
stormy black lime 236
linden honey 40

goat biryani with jackfruit and lime leaf rice 130
loaf cakes 194–5
locust honey 39
loukoumades 170

M
madeleines, brown butter / oak / chestnut 163–5
mango: honey jar ferments 62
manuka honey 16, 40
maple honey: home-smoked honey 65
margarita, smoked mezcal 238
mascarpone, rose roast quince with hazelnut meringue and honeyed 203
mead 30–1
meadow honey 36
meadows 42
meringues 29
  griddled peach and pistachio pavlova with lavender honey 194–6
  rose roast quince with hazelnut meringue and honeyed mascarpone 203
mezcal: mezcal blood orange hibiscus granita 222
  smoked mezcal margarita 238
Michelle Polzine's 10-layer honey cake 207–8
milk: cardamom oats 78
mint: Campari watermelon mint granita 223
  minted lamb chops with herbed grains, pomegranate and sherry vinegar 113–15
miso: garlic fermented honey and miso sourdough 140–3
  Saikyo salmon 116
monofloral honey 36
mushrooms: fennel roast squash with burrata, pickled mushrooms and crispy onions 106
  quick pickled mushrooms 66

sticky roast mushrooms 66
mustard: griddled green mustard salad with smoked almonds 86

N
nectar 42, 45, 46, 48, 57
neonicotinoids 43, 54
Noble, Simon 17
noodles, smoked duck, smashed cucumber, slapped 119–21

O
oak honey 41
  brown butter / oak / chestnut madeleines with orange blossom icing 163–5
  home-smoked honey 65
  sesame Old Fashioned 241
oats, cardamom 78
oilseed rape 26, 36, 42
Old fashioned, sesame 241
olive oil: almond, olive and raspberry loaf cake 184
onions, fennel roast squash with burrata, pickled mushrooms and crispy 106
orange blossom honey 41
  almond briouats 181–3
  baklava 174–5
  blood orange, almond and dark chocolate babka 157
  cardamom Basque cheesecake 200
  halwa chebakia 178–80
  minted lamb chops with herbed grains, pomegranate and sherry vinegar 113–15
  orange, yogurt and cardamom loaf cake 185
  scauratielli 166
orange blossom icing 163–5
oranges: blood orange, almond and dark chocolate babka 157
  mezcal blood orange hibiscus granita 222

orange, yogurt and cardamom loaf cake 185
orchard honey 36, 42

P
panna cotta, fig leaf 210
paprenjacs, cut glass 171
paprika parched peas with butter-roast fennel and garlic green peppercorn mash 110–12
pasta, silky squash 95
pastries: apricot and fennel croissant swirls 144–9
pavlova, griddled peach and pistachio 194–6
peaches: griddled peach and pistachio pavlova 194–6
  peach peppercorn tepache 231
peanuts: honey nut corn cake 197
pears: radicchio pear salad with spiced fritto misto 91
peppercorns: paprika parched peas with butter-roast fennel and garlic green peppercorn mash 110–12
  peach peppercorn tepache 231
peppers, harissa hazelnut beetroot with raw cavolo and dilly red 109
pesticides 43, 54, 57
pheromones 48
pickles: pickled carrots 88
  quick pickled mushrooms 66
pie, lemon pollen 204
pine honey 41
pistachios: griddled peach and pistachio pavlova 194–6
  pollen and pistachio baklava with honeyed hibiscus syrup 174
Plantlife 56
plants: how bees find flowers 48
  pollination 46–9
  recommendations 58–9
plums: double plum ribs and lime leaf slaw 122

pollen 15, 33, 42, 46, 51
pollination 45–9
polyfloral honey 36
pomegranate: minted lamb chops with herbed grains, pomegranate and sherry vinegar 113–15
pork: double plum ribs and lime leaf slaw 122
  honey hong shao rou 136
potatoes: paprika parched peas with butter-roast fennel and garlic green peppercorn mash 110–12
  smoked lime and honey chicken 126
pots 55
prawns (shrimp), guajillo garlic 92
propolis 11, 12
puff pastry: honey tarte tatin 191–3

Q
queen bees 11, 12, 15, 16, 18, 20
quinces: rose roast quince with hazelnut meringue and honeyed mascarpone 203

R
radicchio: charred radicchio and pickled carrot salad with sesame-seared tuna 88
  radicchio pear salad with spiced fritto misto 91
radishes: fermented fennel kimchi 68–9
raspberries: almond, olive and raspberry loaf cake 184
raw honey 26, 35, 36
  apricot caraway tepache 231
  fermented fennel kimchi 68–9
  honey jar ferments 62
  peach peppercorn tepache 231
  preserved lemon focaccia 152–3
Reese, Matt 55
rice, goat biryani with jackfruit and lime leaf 130

ricotta: roasted black sesame ricotta ice cream 216
rose petals: rose roast quince with hazelnut meringue and honeyed mascarpone 203
rose water: pistachio, rose and cardamom babka 156–7
round dance 48
Royal Botanic Gardens Kew 52, 56, 57
rum: stormy black lime 236

S
safflower honey 41
  dark beef and aubergine curry 129
  smoked duck, smashed cucumber, slapped noodles 119–21
  spice cupboard goat with preserved lemons 132–3
saffron and apricot honey buns 154
Saikyo salmon 116
salads: charred radicchio and pickled carrot salad with sesame-seared tuna 88
  griddled green mustard salad with smoked almonds 86
  radicchio pear salad with spiced fritto misto 91
salmon, Saikyo 116
sauce, sour cherry chipotle hot 73
scauratielli 166
Scotch bonnets 31
  honey jar ferments 62
sea lavender honey 41
  Saikyo salmon 116
seedheads 57
'seeding' honey 26
sesame-infused whisky: sesame Old Fashioned 241
sesame seeds: charred radicchio and pickled carrot salad with sesame-seared tuna 88
  halwa chebakia 178–80

roasted black sesame ricotta ice cream 216
sherry vinegar, minted lamb chops with herbed grains, pomegranate and 113–15
silky squash pasta/soup 95
single-origin honey 36
single species honey 36
slaw, lime leaf 122
smoked duck, smashed cucumber, slapped noodles 119–21
smoked honey 30, 65
  gochujang apricot sticky wings 99
  guajillo garlic prawns 92
  sesame Old Fashioned 241
  silky squash pasta/soup 95
  smoked lime and honey chicken 126
  smoked mezcal margarita 238
  sour cherry chipotle hot sauce 73
smoked lime and honey chicken 126
smoked mezcal margarita 238
solitary bees 57
soup, silky squash 95
sour cherry chipotle hot sauce 73
sourdough 33
  garlic fermented honey and miso sourdough 140–3
spice cupboard goat with preserved lemons 132–3
spring 11–15
squash: fennel roast squash with burrata, pickled mushrooms and crispy onions 106
  silky squash pasta/soup 95
sticky roast mushrooms 66
stormy black lime 236
sucrose 29
summer 16–19
supermarket honey 26
swarms 15, 16, 18
sweets: scauratielli 166

INDEX 249

T

tacos: achiote orange tacos with jackfruit and cauliflower 100
  honeyed chipotle lamb tacos 103
tarts: custard tart! 188–90
  honey tarte tatin 191–3
  lemon pollen pie 204
tea: fiery ginger kombucha 228
Tej 31
tepache: apricot caraway tepache 231
  peach peppercorn tepache 231
  tepache Bourbon sour 232
terroir 32–43
thixotropic honey 16
thyme honey 41
  almond, olive and raspberry loaf cake 184
  fennel roast squash with burrata, pickled mushrooms and crispy onions 106
tomatoes: black garlic and lime tomatoes 85
  dark beef and aubergine curry 129
tortillas: achiote orange tacos with jackfruit and cauliflower 100
  honeyed chipotle lamb tacos 103
trees 51, 58
tuna: charred radicchio and pickled carrot salad with sesame-seared tuna 88
tupelo honey 41
  double plum ribs and lime leaf slaw 122
  honey nut corn cake 197
  honeyed chipotle lamb tacos 103

V

varietals 39–41
vodka: eucalyptus honey granita 222

W

waggle dance 48
walnuts: almond, walnut and stem ginger baklava with orange blossom honey syrup 175
  coffee and walnut cake 209
  cut glass paprenjacs 171
  loukoumades 170
warqa: almond briouats 181–3
watermelon: Campari watermelon mint granita 223
weeding 56–7
whisky: sesame Old Fashioned 241
wildflower honey 36, 41, 42
  burnt honey dulce ice cream 219
  fiery ginger kombucha 228
  Michelle Polzine's 10-layer honey cake 207–8
  pistachio, rose and cardamom babka 156–7
  radicchio pear salad with spiced fritto misto 91
wildflowers 51
winter 23
worker bees 11, 12, 15, 16, 33, 35, 45, 46, 48
Wyndham Lewis, Sarah 51

Y

yogurt: griddled peach and pistachio pavlova with lavender honey 194–6
  harissa hazelnut beetroot with raw cavolo and dilly red peppers 109
  minted lamb chops with herbed grains, pomegranate and sherry vinegar 113–15
  orange, yogurt and cardamom loaf cake 185
  smoked lime and honey chicken 126
Yucatan honey 41
  achiote orange tacos with jackfruit and cauliflower 100
  apricot caraway tepache 231
  harissa buttermilk dressing 72
  harissa hazelnut beetroot with raw cavolo and dilly red peppers 109
  mezcal blood orange hibiscus granita 222
  peach peppercorn tepache 231

Thank You

This book would not exist without my marvellous editor Harriet Webster. Thank you Harry, for not only giving me this opportunity, but championing my efforts and ideas every step of the way. Thank you Emily Lapworth, Kim Lightbody, Anna Wilkins, Tamara Vos, Emma Cantlay, Charlotte Whatcott and Florence Blair for your incredible work. Thank you Mark Diacono, for your words of wisdom and encouragement, and to Emily Sweet, for seeing my potential.

Thank you to Michelle Polzine, Christine Benlafquih and Helen Graves for allowing me to share your wonderful recipes in this book, and to Quality Chop House, Yotam Ottolenghi, Ixta Belfrage, Dishoom, Claire Ptak, Flor, Honey & Co, Popham's, Raymond Blanc, Richard Godwin and Santo Borzi for influencing my cooking with your joyful creations.

Thank you Simon and Chloe, for generously welcoming me into the new forest and introducing me to your wonderful bees. Thank you Zena, for commissioning me to write about Simon's bees and championing good honey.

HH, thank you for throwing yourself behind this adventure, as chief taster and supporter of my dreams. Thank you for giving over much of your freezer to 101 ice cream tests and often waiting until nearly midnight to eat dinner...

Thank you to my mum Louise, and grandmothers Mavis and Betty, for raising me in the kitchen. You gave me my love of cooking.

Author Biography

Amy Newsome is a horticulturist and garden designer, food writer and beekeeper based in London, UK. After a career in fashion marketing, she retrained in gardening and beekeeping, working for Raymond Blanc, organic grower Anna Greenland and the Royal Botanic Gardens, Kew. She has also worked with prison reform charity Food Behind Bars, helping to bring bees and kitchen gardening into prisons. She now spends her time designing and growing gardens, beekeeping, writing, and cooking her socks off – preferably in the garden. She has written for multiple publications, including *Bloom* magazine, on subjects such as single-origin honey and cooking over fire.

Managing Director *Sarah Lavelle*

Commissioning Editor *Harriet Webster*

Copy Editor *Sally Somers*

Art Director and Designer *Emily Lapworth*

Photographer *Kim Lightbody*

Food Stylist *Tamara Vos*

Prop Stylist *Anna Wilkins*

Head of Production *Stephen Lang*

Production Controller *Gary Hayes*

First published in 2023 by Quadrille,
an imprint of Hardie Grant Publishing

Quadrille
52–54 Southwark Street
London SE1 1UN
quadrille.com

Text © Amy Newsome 2023
Photography © Kim Lightbody 2023
Design and layout © Quadrille 2023

All rights reserved. No part of the book may be reproduced, stored in a retrieval system, or transmitted in any form or by any means, electronic, electrostatic, magnetic tape, mechanical, photocopying, recording or otherwise, without the prior permission in writing of the publisher.

The rights of Amy Newsome to be identified as the author of this work have been asserted by her in accordance with the Copyright, Design and Patents Act 1988.

Cataloguing in Publication Data: a catalogue record for this book is available from the British Library.

ISBN: 978 1 78713 943 5

Printed in China

Many thanks to the women who generously allowed me to share their recipes in this book:

Flatbreads recipe on page 132 adapted from *Live Fire* by Helen Graves, copyright © 2022. Reprinted by permission of Hardie Grant Books.

Almond briouats recipe on page 181 and Halwa Chebakia recipe on page 178 by Christine Benlafquih, Taste of Maroc (tasteofmaroc.com)

10 layer honey cake recipe on page 207 adapted from *Baking at the 20th Century Cafe* by Michelle Polzine, copyright © 2020. Reprinted by permission of Artisan, an imprint of Hachette Book Group, Inc.

Cook's notes:

All honey varietals suggested in the recipes are merely a guide. Do use what you have to hand if you prefer, and look for what types of honey are local to you. Where possible, please try to source raw (minimally filtered, unheated) local honey.

All eggs are medium free range.

All citrus fruit is unwaxed.

Please use the best quality vegetables, meat and fish you can afford, ideally organic.

See souschef.co.uk for any unusual ingredients.

For aspiring beekeepers:

If you are interested in learning more about bees or beekeeping, search for the beekeeping association in your country. In the UK this is bbka.org.uk and in the US abfnet.org.